The Memory Phenomenon in Contemporary Historical Writing

Patrick H. Hutton

The Memory Phenomenon in Contemporary Historical Writing

How the Interest in Memory Has Influenced Our Understanding of History

Patrick H. Hutton
University of Vermont
Burlington, Vermont, USA

ISBN 978-1-137-49464-1 (hardcover)   ISBN 978-1-137-49466-5 (eBook)
ISBN 978-1-349-69737-3 (softcover)
DOI 10.1057/978-1-137-49466-5

Library of Congress Control Number: 2016939235

Cover image © Superstock / Alamy Stock Photo

Printed on acid-free paper

This Palgrave Macmillan imprint is published by Springer Nature
The registered company is Nature America Inc. New York

For Jennifer, Sean, Matthew, Scott, and Jeffrey

Personal Acknowledgments

My book is the culmination of a decade of research and writing about the relationship between memory and history. I have written a variety of articles, essays, and reviews that have made it possible for me to attempt this synthesis, and I have derived great satisfaction in drawing together ideas, insights, and perspectives of the able scholars in this burgeoning field of scholarship, among the most important in the historiography of the late twentieth century. Along the way, I have appreciated the advice and encouragement of editors, scholars, and friends. I thank Lee Brown, Rachel Fuchs, Trever Hagen, Siobhan Kattago, Daniel Levy, Fiona Mcintosh-Varjabédian, Linda Mitchell, Jeffrey Olick, Nancy Partner, Mark Phillips, Sabine Schindler, Natalia Starostina, Richard Sugarman, and Anna Lisa Tota. I also thank the Humanities Institute at Arizona State University and the Retired Professors Association at the University of Vermont for grants that furthered my research and writing.

Acknowledgments

Earlier versions of material in my book have appeared in learned journals or as book chapters: "Pierre Nora's *Les Lieux de mémoire* Thirty Years After," *Routledge International Handbook of Memory Studies*, ed. Anna Lisa Tota and Trever Hagen (London: Routledge, 2015), 28–40 (published with permission of the Taylor & Francis Group); "Pioneering Scholarship on the Uses of Mythology in the Remembrance of Modern Wars," *Between Memory and Mythology*, ed. Natalia Starostina (Newcastle upon Tyne: Cambridge Scholars Publishing, 2015), i–xvi (published with permission of Cambridge Scholars Publishing); "Reconsiderations of the Idea of Nostalgia in Contemporary Historical Writing," *Historical Reflections* 39/3 (2013): 1–9 (published with permission of the editor); "Memory: Witness, Experience, Collective Meaning," *The Sage Handbook of Historical Theory*, ed. Sara Foot and Nancy Partner (London: Sage Publications, 2013), 354–377. I thank the editors and publishers for permission to republish this material here.

Contents

CHAPTER 1

Historiographical Background to the Memory Phenomenon

From the History of an Ancient Idea into the Historiography of a Contemporary Vogue

This book is a reflection on my peregrinations in memory studies, and offers an overview of the remarkable historical interest in the topic of collective memory since the late 1970s. Some 20 years ago I published *History as an Art of Memory* (1993).[1] That book was a study in the history of ideas. I explored the way the ancient art of memory was reinvented in modern times within the context of philology, romantic poetry, depth psychology, and historiography. The English cultural historian Frances Yates served as my intellectual guide. As an early contributor to the study of the relationship between collective memory and history, I sometimes strayed into the middle ground between the two. At the time, some scholars misconstrued my purpose, and claimed that I was eliding them.[2] So let me be clear at the outset about my understanding of their relationship. History and memory share a common curiosity about the past. Though they may at times overlap as perspectives of the present on the past, they are different in their resources and their contributions to culture. History is rational and analytical; memory is emotional and inspirational. Moreover, their

---

[1] (Hanover, NH: University Press of New England, 1993).
[2] Notably Dominick LaCapra, *History and Memory after Auschwitz* (Ithaca: Cornell University Press, 1998), 23–26, who took a sentimental autobiographical note in my preface to be the thesis of my book, and as such the key to my unconscious intent in writing it.

© The Editor(s) (if applicable) and The Author(s) 2016
P.H. Hutton, *The Memory Phenomenon in Contemporary Historical Writing*, DOI 10.1057/978-1-137-49466-5_1

appeal to the past is different. History fixes the past in a narrative that aspires to provide a measure of certainty about what the past was like, but always at a critical distance. Memory, by contrast, may at any moment evoke the past in all of its possibilities, importing past into present insofar as that might be imagined. As philosopher Paul Ricoeur remarked, memory is a "little miracle" in its resources for creativity. In this respect, it may inspire the historian, too.[3]

This book, by contrast, is primarily about the historiography of the scholars' inquiry into the relationship between collective memory and the rhetoric of historical conceptualization during the late twentieth century. For historians, the topic of memory appeared to emerge precipitously within the scholarship of the late 1970s.[4] A marginal, somewhat arcane interest within the history of ideas during the 1960s—notably through Frances Yates's highly acclaimed study of the Renaissance art of memory—memory studies by the turn of the twenty-first century had reshaped the research and understanding of cultural history, enriching both its methods and content.[5] Scholarly discourse on the topic of memory quickened during the 1990s as varied approaches converged, gathering force in the volume and array of subject matter in a hyperbolic ascent into what came to be characterized as memory studies by the turn of the twenty-first century. As a new arena of historical investigation that matured rapidly, the phenomenon of memory studies sheds light on the way a field of historiography develops—from bold pioneers blocking out new interpretations, to more discerning specialists who follow, before moving on to appreciative latecomers who take research in new directions as the interpretative insights of the pioneers begin to fade from view. The historiography of memory studies also reveals the way in which initially provocative interpretative forays into a new field of scholarly inquiry are eventually

---

[3] In this distinction, I follow Paul Ricoeur, *La Mémoire, l'histoire, l'oubli* (Paris: Seuil, 2000), 644. See my essay, "Memory," in the *New Dictionary of the History of Ideas*, ed. Maryanne Horowitz (Detroit, MI: Thomson/Gale, 2005), 4: 1418–22.

[4] For perspectives on the rise of memory studies, see Kerwin Klein, "On the Emergence of Memory in Historical Discourse," *Representations* 69 (2000), 127–50; Chris Lorenz, "Unstuck in Time. Or, The Sudden Presence of the Past," in *Performing the Past; Memory, History, and Identity in Modern Europe,*, ed. Karin Tilmins, Frank van Vree and Jay Winter (Amsterdam, Netherlands: University of Amsterdam, 2010), 67–102; Wulf Kansteiner, "Finding Meaning in Memory: A Methodological Critique of Collective Memory Studies," *History and Theory* 41 (May 2002): 179–197. For an overview of the field, see Geoffrey Cubitt, *History and Memory* (Manchester, UK: Manchester University Press, 2007); Astrid Erll, *Memory in Culture* (Palgrave Macmillan, 2011).

[5] Yates, *The Art of Memory* (Chicago: University of Chicago Press, 1966).

reassessed and integrated into a larger body of scholarship. By the 2010s, memory studies had become an interdisciplinary venture, loosening its ties to the historiographical movement of the 1970s out of whose matrix it had emerged.

In framing my study, I address the questions: why so much interest in memory among historians, and why did it emerge in the late twentieth century? I consider them in two contexts: one historical, the other historiographical:

LATE TWENTIETH-CENTURY HISTORY: A CRISIS OF IDENTITY

From a historical perspective, the historians' preoccupation with memory in the late twentieth century may be attributed to anxieties about the breakdown of long-standing collective identities undermined by new historical realities that contributed to their dissolution. In the post-World War II era, particularly by the 1970s, new realities had emerged to undercut the modern historical narrative, indeed to render it irrelevant. Globalizing economic forces challenged the primacy of national identity. A new economy of consumerist desire displaced the older one of human need. The fads of consumerism drove fantasies that blurred the line between real and vicarious identities. The distinction between high and popular culture dissolved in the face of a consumerist culture that promoted an abundance of homogenized material riches for those who could afford them, while relegating the workers who produced them in the far corners of the world to endemic poverty. The long twentieth-century struggle for women's rights and opportunities played into rethinking the nature of gender identity itself by century's end. Most imposing of all was a revolution in technologies of communication whose accelerating pace eclipsed typographic culture. New media altered ways of organizing knowledge, exporting vastly expanding realms of data to readily accessible electronic archives, with far-reaching implications for what and how we remember. Learning in a digital age was transformed, especially for the young, to such a degree that computer scientists speculated about an eventual convergence of biological and artificial intelligence.[6] In a world whose culture

---

[6] See the prophecy by Ray Kurzweil, *The Singularity is Near; When Humans Transcend Biology* (New York: Penguin, 2005), as well as the skeptical critique by Nicholas Carr, *The Shallows; What the Internet is Doing to Our Brains* (New York: Norton, 2011), esp. 175–76.

was being reconfigured in so many ways, historians would begin to rethink the meaning of collective identity in the globalizing culture of the contemporary age. Memory, the seat of such knowledge at all levels of human experience, would rise up to meet their inquiries, inspiring them to think about the past in relation to the present in innovative ways. Over time, collective memory, conspicuously identified with the commemorative rituals of the nation-state, would break free of the constraints of that association to reveal a myriad of particular identities in global settings, mirroring the changing realities of the late twentieth century.

It was not just the unsettled present, but also a past full of haunting memories that troubled historians about the grand narrative of modern history. Old and unresolved problems raised new questions about the historical meaning of the twentieth century in light of the massive death and destruction that it had witnessed. Two world wars, a devastating economic depression in the era between them, the calculated genocide of European Jews, and the American use of the atomic bomb as a weapon of war dispelled any and all notions that the twentieth century had bequeathed to the present age a historically intelligible route toward the making of a better world. The atrocities of the Holocaust, far from receding into the past, loomed larger with the passage of time as an unrequited memory of reality that defied comprehension. What was one to make of sublime evil committed by the Nazi government of a once enlightened nation in a historical age supposedly advancing the human condition? The debates of the "Historians' Dispute" among German scholars during the 1980s underscored their awareness that the old narrative of history was no context in which to interpret the historical meaning of the conscious plan to exterminate a specific group of people solely for its genetic inheritance. These were recognized as crimes against humanity, a past whose meaning had yet to be mastered by historians.[7] The power of trauma to block remembrance became the focus of their scholarly research. As method, the psychoanalytic theory of Sigmund Freud, banished with the stalled venture of psychohistory during the 1960s, came to the fore once more in this avenue of scholarship.[8]

---

[7] For an overview of the dispute, see Charles S. Maier, *The Unmasterable Past; History, Holocaust, and German National Identity* (Cambridge, MA: Harvard University Press, 1988).

[8] For the bridge between psychohistory and renewed interest in Freud in memory studies, see Saul Friedländer, *History and Psychoanalysis* (1975; New York: Holmes & Meier, 1980), esp. 9–42.

LATE TWENTIETH-CENTURY HISTORIOGRAPHY: A CRISIS OF METANARRATIVE

From a historiographical perspective, the memory phenomenon in late twentieth-century historiography may be construed as the first serious effort to assess the relationship between memory and history. For much of the nineteenth century, historians, like their readers, thought little about their differences, and tended to conflate them in their excursions into the past. They aspired not only to explain the realities of those times but also to convey to their readers some feeling for its imagination. The public came to value the study of history not only for intellectual edification but also for emotional empathy. Long after their work has been superseded by more exacting scholarship, well-known historians such as Jules Michelet and Benedetto Croce continued to be admired for their capacity to evoke the passion in the pageant of the past. Memory and history were thought to cooperate in the quest to approach the impossible dream of bringing the past to life once again. Sympathy for this interplay of memory and history would surface once more in memory studies toward the turn of the twenty-first century, this time from a critical rather than a naive perspective.

The professionalization of historical scholarship of the late nineteenth century, however, put its accent on their opposition. Memory and history were understood to operate in tandem. History offered itself as the official form of memory. It claimed to provide a rigorously critical interpretation of the remembered past, chastening collective memory by deflating its exaggerations and excising its misconceptions. It prided itself on its accuracy, objectivity, dispassion, and critical distance from the past. It confirmed that claim by its appeal to method and to evidence. Historical scholarship was regarded as a high responsibility because it corrects the misperceptions of memory, and so lends stability to human understanding of the past. In its best analyses, history in its modern scholarly guise offered a perspective on the past based on reliable certainties, and so was characterized as a particular kind of science. As French historian Jacques Le Goff put it, "Memory is the raw material of history." History begins where memory ends. Its authority depends on the historicist proposition that there is an underlying temporal foundation in which all past experience is grounded. The timeline of history serves as the essential frame of reference for a universal "science of time."[9]

---

[9] Jacques Le Goff, *History and Memory* (New York: Columbia University Press, 1992), xi, 214.

By the late 1970s, though, this simple formula for explaining memory's subordination to history had come to be recognized as inadequate. It is in this context that historian Pierre Nora published his *Lieux de mémoire* (1984–1992), an ambitious collaborative study of the mnemonic sources of the French national identity as they had sprung forth since the Middle Ages. The standard narrative of modern French history that had served for more than a century as the framework for historical scholarship had lost its power of appeal for practicing historians. Meanwhile, the interest in collective memory was surging, notably in studies of commemorative practices.[10] Such scholarship revealed that there were many ways in which memory and history were intertwined. Following the initiative launched by Nora and his colleagues, three principal lines of inquiry into the puzzles of memory's relationship to history came to the fore during the crucial decade of the 1980s, not only in France but throughout Europe and North America: the politics of commemorative practices; the cultural implications of the transition from oral to literate cultures; the disabling effects of trauma on historical understanding, with particular emphasis upon the Holocaust of European Jews during World War II. These pathways would guide directions of historical scholarship on the memory phenomenon until the turn of the twenty-first century.

Symptomatic of the crisis that precipitated the memory phenomenon was the breakdown of the "grand narrative" of modern history, a proposition advanced by French philosopher Jean-François Lyotard in a book about the "postmodern condition."[11] Lyotard argued that the narrative of the rise of Western Civilization as the vehicle of reform in the name of the modern imperative of progress, both economically (as greater and more equitably distributed prosperity) and morally (as civic purpose and responsibility) had lost its conceptual power to frame historical understanding. The paradigm for such writing had been born of the European Enlightenment and confirmed by the vast institutional upheaval ushered in by the French Revolution. These intellectual and political forces fostered expectations of the modernizing role of the emerging nation-state, while showcasing the bourgeoisie as the entrepreneurial elite that would drive the new urban industrial economy, reshape politics around

---

[10] Exemplary is John R. Gillis, ed., *Commemorations; The Politics of National Identity* (Princeton: Princeton University Press, 1994).

[11] Jean-François Lyotard, *La Condition postmoderne* (Paris; Editions de minuit, 1979), 29–35.

ideological imperatives, and refashion the high culture of science, the arts, and literature. Its bias would engender class struggle and imperialist ventures abroad. By the late nineteenth century, Europe had colonized much of Africa and Asia, politically and culturally. Libertarian in its conceptualization, the grand narrative was democratic in its moral intentions. It spoke to the beliefs and to the needs of the left-of-center statesmen of nineteenth-century Europe and America.

Within twentieth-century European historiography, however, this foundational narrative of the rise of modern civilization under the aegis of the nation-state had come to be rivaled by alternative metanarratives—that of Marxism among left-wing scholars for its political commitments and that of the Annales movement among more erudite historians devoted to widening the sphere of archival research.[12] Both of these historiographical movements turned to social, economic, and cultural topics that conventional historians had once ignored, and they emphasized the hidden power of the impersonal workings of historical forces relentlessly imposing the past upon the present. Marxism had gained force in late nineteenth-century Europe as a fighting philosophy for the labor movement.[13] After World War I, it had been co-opted by Soviet Bolshevism to become a shibboleth for the omnicompetent state in the Soviet Union, a rationale for its policies for the better part of the twentieth century. Meanwhile Marxism as a critical philosophy of history continued to fascinate Western European intellectuals.[14] It may have lost the allure of its Metahistorical claims. But it continued to exercise an enduring appeal as a philosophy of praxis, an investigative tool in the service of consciousness raising, for it professed to illuminate the deep economic structures of historical reality hidden beneath political and cultural illusions. Marxism in this guise had taken on new life after World War II, thanks to the role of Communists in the resistance movements that fought fascism across Europe. It held a particular mystique for French intellectuals coming of age in the postwar

[12] Philip Daileader and Philip Whalen, "Introduction: The Professionalization of the French Historical Profession," *French Historians 1900–2000*, ed. Daileader and Whalen (Chichester, UK: Wiley-Blackwell, 2010), xix–xxiv; Guy Bourdé and Hervé Martin, *Les Ecoles historiques* (Paris: Seuil, 1983), 245–306.

[13] George Lichtheim, *Marxism; An Historical and Critical Study* (New York: Columbia University Press, 1961), 216–17,222–33.

[14] Georg G. Iggers, *Historiography in the Twentieth Century; From Scientific Objectivity to the Postmodern Challenge* (Hanover, NH: University Press of New England, 1997), 85–94.

era.[15] In academic circles, it exercised significant influence among historians of the French Revolution and the historiographical tradition that followed from it. In addition to providing an explanation of the role of powerful economic forces underpinning historical change, it offered a direction of moral intention for building a more just and egalitarian society, according to Georges Lefebvre, its most venerated scholar.[16] By the 1970s, however, the Marxist theory of history had grown stale in its reiteration, and many of these intellectuals expressed disenchantment with its constraining paradigm of interpretation, not to mention the waning of their enthusiasm for Communist politics. As historian François Furet, himself a former adherent of the French Communist Party, remarked, it had become impossible to disassociate Marxism in the twentieth century from its embodiment in Soviet communism.[17] In confessional style, some repudiated their youthful allegiance to the Party and more generally the determinism implicit in Marxist theory.[18]

In postwar Germany, too, scholarly enthusiasm for the Marxist-inspired Frankfurt school of social criticism, launched by Max Horkheimer and Theodor Adorno during the interwar years, was waning by the 1970s. While abandoning Marx's teleological theory of history, scholars in this tradition had remained committed to his method, based on the kind of rational critical analysis that he had pioneered.[19] For English scholar Paul Connerton, the Frankfurt style pursuit of the "dialectics of enlightenment" as a historical perspective had lost touch with the new social realities of the late twentieth century. Its leading philosophers, he argued, had sacrificed practical insight to "an enveloping orgy of abstractions" of diminishing

---

[15] George Lichtheim, *Marxism in Modern France* (New York: Columbia University Press, 1966), 80–89; François Furet, *Lies, Passions, and Illusions* (Chicago: University of Chicago Press, 2014), 34–35.

[16] Georges Lefebvre, *The Coming of the French Revolution* (Princeton NJ: Princeton University Press, 1947), 217–20.

[17] François Furet, *Le Passé d'une illusion* (Paris: Robert Laffont, 1995), 7–13.

[18] Mona Ozouf, Jacques Revel, and Pierre Rosanvallon, eds. *Histoire de la Révolution et la révolution dans l'histoire: entretiens avec François Furet* (Paris: AREHESS, 1994), 4–8; Emmanuel Le Roy Ladurie, *Paris/Montpellier; P.C.-P.S.U., 1945–1963* (Paris: Gallimard, 1982).

[19] Martin Jay, *The Dialectical Imagination; A History of the Frankfurt School and the Institute of Social Research, 1923–1950* (Boston: Little Brown, 1973), 253–80; idem, *Marxism and Totality; The Adventures of a Concept from Lukács to Habermas* (Berkeley: University of California Press, 1984), 1–20.

appeal to a scholarly following.[20] While finding their way into other intel-lectual movements, some disenchanted Marxists, nonetheless, maintained a sentimental attachment to its heritage. Philosophical celebrity Jacques Derrida, as late as the 1990s, made a belated case for the afterlife of a "ghostly Marxism."[21] In his way, he was mourning the passing of a philos-ophy that had animated the youth of his generation. Marxism had always augured the future in interpreting the past, but its reading of that past was now sliding into irrelevance.

Annales scholarship was another story in the search for an alternative to the metanarratives of national history. It, too, aspired to a broadly conceived overview, a "total" history that traced the storylines of economic, social, and environmental forces while downplaying politics.[22] Its leading histori-ans offered a sophisticated theory of deep structures of history, whose forms changed according to a tempo of time that was slow, sometimes practically immobile.[23] While repudiating the notion of the teleological unfolding of patterns of history, Annalistes, nonetheless, based their research on quanti-tative techniques calculated to reveal the determining power of vast imper-sonal historical forces, whose influence was fully revealed only when the serial patterns of the past were considered *à la longue durée*. The Annales movement acquired prestige among historians everywhere for the widen-ing horizons of scholarship that it opened for research.[24] But after three generations of work within this scholarly tradition, the Annales paradigm, too, had lost the fervor of the movement's founders in the 1920s. The ambitions of the Annalistes had exceeded their conceptual reach toward synthesis based on empirical findings. The lodestar governing their pur-suits in the agenda set by the movement's founders had grown dim amidst the pluralism of well-researched, discrete studies carried out in its name. What unity it possessed by the 1980s resided in the network of its most

[20] Paul Connerton, *The Tragedy of Enlightenment; An Essay on the Frankfurt School* (Cambridge,, UK: Cambridge University Press, 1980), 134.

[21] Jacques Derrida, *Specters of Marx* (London: Routledge, 1994), 13–18.

[22] For the viewpoints of pioneers of the Annales movement, see the collections of essays by Lucien Febvre, *Combats pour l'histoire* (Paris: Armand Colin, 1992); Fernand Braudel, *Ecrits sur l'histoire* (Paris; Flammarion, 1969).

[23] Emmanuel Le Roy Ladurie, "L'Histoire immobile," *Le Territoire de l'historien* (Paris: Gallimard, 1978), 2: 7–34.

[24] For an overview, François Dosse, *L'Histoire en miettes: Des "Annales" à la "nouvelle his-toire"* (Paris: La Découverte, 1987), 212–59; Stuart Clark, ed., *The Annales School: Critical Assessments* (London: Routledge, 1999).

prestigious scholars rather than an agenda for investigation. Indicative was the retreat of some of its leading scholars into more personalized accounts of their path into history, a phenomenon labeled *égo-histoire* by the 1980s.[25]

The paradox was that in the midst of the breakup of the grand narrative, all sorts of new approaches to history were presenting themselves, each begging for a narrative of its own. In this respect, the decade of the 1960s might be regarded as a golden age of historiography for the new directions of historical research pursued by a younger generation of scholars in Europe and America. That decade witnessed an explosion of new subject matter: women's history, global history, post-colonial history, historical psychology, African-American history, as well as histories of an array of minority groups.[26] This pluralistic turn in historiography is hardly surprising. All of these topics called for a reexamination of the past in light of the way the culture, and more specifically the newly conceived notion of a culture of politics, was being refashioned in the present age.[27]

New historical interests, together with old and unrequited memories, thus contributed to the reorientation of the historians' perspective in emerging networks of historical scholarship around the globe during the 1970s. Whereas historians had once favored continuity between past and present, now they remarked upon disruptions between them; whereas they had previously looked forward with great expectations of the future, now they looked back upon the failures of the near past of the twentieth century. The task, then, was not so much to revise standard narratives, as historians who came of age during the 1960s counseled, but rather to discard them so as to look once more at the memories that had initially inspired them.[28] Deeper than particular attempts at metanarrative was an emerging skepticism about the historical determinism that they implied. Furet expressed the sentiment well. If one reviews twentieth-century history, he

[25] Pierre Nora, ed., *Essais d'égo histoire*. Paris: Gallimard, 1987.

[26] A perspective on the 1960s as a golden age of historiography is found in the essays contributed to Felix Gilbert and Stephen Graubard, eds., *Historical Studies Today* (New York: Norton, 1972).

[27] The Cold War was still a framework for defining the history of the post-World War II era; but its framework of interpreting new historical forces, especially those of a social and cultural nature, seemed limited. Among younger scholars, diplomatic history was coming to be considered a backwater of historical scholarship during the 1970s. See Charles S. Maier, "Making Time: The Historiography of International Relations," in *The Past Before Us*, ed. Michael Kammen (Ithaca: Cornell University Press, 1980), 355–56.

[28] See the discerning discussion by world historian William McNeil, *Mythhistory and Other Essays* (Chicago: University of Chicago Press, 1986), 3–42.

contended, it is its visible personalities and its historical contingencies that explain the course of events, not some hidden mechanism to be revealed by the cognoscenti.[29] Historians were returning from their fascination with *la longue durée* to reevaluate the past within the present, giving greater attention to historical contingencies that had altered the anticipated course of history. This turn provided an opening for the recall of neglected historical experience, and as such was a point of entry for the reevaluation of the relationship between memory and history.

THE MEMORY PHENOMENON IN RELATION
TO POSTMODERN HISTORIOGRAPHY

A deeper philosophical proposition about the nature of history was also at issue. The search for synthesis in a grand metanarrative had been based upon the long-standing theory of historicism, by which a timeline of human experience was understood to serve as the unifying ground of historical interpretation.[30] If the notion of history as the storyline of the rise of the West had lost its meaning for the present age, so too had faith that the relationship between past and present could be explained in terms of a backbone narrative emerging out of the depths of time. There had been too much displacement, destruction, and death in the wars and economic crises of the twentieth century to contend that somehow all of these disruptive forces might be adapted to a framework of history as an ongoing and uplifting journey.[31] Too many unanticipated misfortunes had intervened to permit the plotting of modern history as a saga from recognizable beginnings toward an expectant end. Having discarded the historicist narrative of modern history, historians asked: how might they begin to reassess the historical meaning of the present age?

Some scholars answered the question in political terms. They saw the denouement of a half century of ideological rivalry, orchestrated by the superpower USA and the Soviet Union, as the apparent triumph of liberal democracy over its collectivist rival. New and sometimes strange

---

[29] François Furet, *Le Passé d'une illusion*, 773–809; idem, *Lies, Passions, and Illusions* (Chicago: University of Chicago Press, 2014), 34, 39–42.

[30] Georg G. Iggers, "Historicism: The History and Meaning of the Term," *Journal of the History of Ideas* 56 (1995), 142–51.

[31] For the scope of the destructions of the world wars of the twentieth century, see Mark Mazower, *Dark Continent; Europe's Twentieth Century* (New York: Random House, 1998), 212–13.

prophecies about the future of history were voiced in the wake of the revolution of 1989 in Eastern Europe, the collapse of the Soviet Union, and the conclusion of the Cold War. Historians Francis Fukuyama and Lutz Niethammer advanced the notion that humankind had arrived at the "end" of history and was entering a "posthistorical" age. Both meant to be provocative rather than literal in these proclamations. We have asked too much of the idea of history, they contended, and it is better to appreciate modern history with more modest expectations about what the future holds. Both argued that what we have called history since its professionalization in the late nineteenth century was in fact the story of the bourgeoisie in its efforts to refashion civilization in its own image through the instrument of the nation-state. Both prophesied the coming of a time of greater political harmony, metaphorically a biblical "peaceable kingdom" in which the struggles of the modern age would give way to the managerial solutions of another about to come. To be fair, Fukuyama was making a philosophical as much as a historical argument. He saw our times as one in which liberal democracy had come to be recognized as a moral imperative beyond which humankind cannot go in pursuit of the good society.[32] Niethammer was resurrecting forgotten nineteenth-century theorists who had forecast the coming of planned societies.[33] From our perspective in the deeply troubled early decades of the twenty-first century, the future these theorists envisioned seems utopian, their commentary of value more as critique than expectation.

The discourse of Fukuyama and Niethammer about the end of modernity prepared the way for rethinking contemporary historiography in light of an emerging discussion about "postmodern" culture. This term "postmodern" has no settled definition, for it has meant many things to many scholars over the course of the late twentieth century. It emerged as a neologism among art historians as early as the 1950s. Art historian Charles Jencks explains how artists disassembled the architectural structures of modernity into their component parts and then reassembled them in incongruent, surprising, and often provocative ways.[34] Later in the century, the term was taken up by literary critics and eventually by

---

[32] Francis Fukuyama, *The End of History and the Last Man* (New York: Avon Books, 1993), xi.

[33] Lutz Niethammer, *Posthistoire; Has History Come to an End?* (London: Verso, 1992), 7–19.

[34] Charles Jencks, *What is Post-Modernism?* (Chichester, UK: Wiley, 1996), 29–40.

historians.[35] Postmodernism from the historians' perspective has many connections with the deconstruction movement among literary theorists, notably in their interest in the rhetorical forms out of which narratives are constructed. Historian Perry Anderson was among the first to sketch the implications of postmodern theory for interpreting popular culture in the contemporary age. Following the work of literary critic Fredric Jameson, he mapped the landscape of postmodern totems of the new consumerism, which despite dismissal by intellectually sophisticated critics for its mundane imagery, nonetheless, spoke to the cultural perceptions of the present age. Anderson emphasized the transformative power of late capitalism to promote cultural fantasies in the fabrication of a new economy of desire. Corporate entrepreneurs took advantage of the rising power of new media, which enlisted memory in its image making, remodeled to suit the needs of consumerist advertising. The effect was to blur the line between reality and fantasy, to remake culture in a consumerist image.[36]

Postmodernism as a discourse of the 1970s stimulated interest in the figuration of the historians' narratives. Thanks especially to Hayden White's *Metahistory* (1973), historians became more sensitive than ever before to the way the rhetoric of historical narrative shapes the meaning we find in the past. White showed how content is inextricably bound up with its textual presentation, making clear the way narrative shapes historical understanding. Historical narratives, therefore, might with profit be disassembled into their component parts to better understand their makeup in their bias, intentions, and most importantly strategies of presentation. No history can escape the shape of its representation.[37]

While contributing to the breakup of the grand narrative, however, deconstruction did little to promote the construction of a new one. One could demonstrate how the old narrative was timeworn and no longer addressed the realities of the present age. But the absence of an overarching temporal framework redirected historians to the places of memory out of which the narrative of modern historiography had first emerged. In this

---

[35] For an overview of the conceptualization of postmodernism, see Andreas Huyssen, *After the Great Divide: Modernism, Mass Culture, Postmodernism* (Bloomington: Indiana University Press, 1987), 178–221.

[36] Perry Anderson, *The Origins of Postmodernity* (London: Verso, 1998), esp. 47–77; see also Fredric Jameson, *Postmodernism or, the Cultural Logic of Late Capitalism* (Durham NC: Duke University Press, 1991), 1–54.

[37] Hayden White, *Figural Realism; Studies in the Mimesis Effect* (Baltimore: Johns Hopkins University Press, 1999), 8–13, 53–54, 88–91.

way, postmodern discourse set preconditions for the rise of the memory phenomenon of the 1980s. While the term postmodern defies explicit definition, it remains in common usage among historians to characterize the present age as a time in history apart from the modern age, and one that has yet to find its bearings.[38]

An intriguing perspective on postmodern historiography was offered by scholars who addressed the issue of changing ways in which historical time has been ordered across the ages. Launched as a discussion of the "semantics" of the modern conception of historical time by German historian Reinhart Koselleck, this line of argument was followed up comprehensively by French historian François Hartog in his thesis about changing "regimes of historicity," by which he means conceptual frameworks for understanding historical time. He proposed that we direct our attention to the ways in which historians have reconceived the nature of historical time across the ages. Hartog posited three such regimes: *historia magistra vitae*, developed out of antiquity; modernity from the eighteenth century; the present age. Each privileged a differed moment of historical time: *historia magistra vitae* an exemplary golden age in the past; modernity an expectant future; presentism a preoccupation with the now time of history. Ours, he argues, is such a time in history.[39]

Hartog advanced two key arguments to illustrate his theory of the present as a point of departure for evaluating the meaning of the past. His first concerned the gathering appeal over the course of the twentieth century of the idea of the acceleration of time. With ever greater speed, the force of innovation was eroding the inertial power of the past, creating the perception that the heritage of the past was slipping into obscurity in the face of present needs. The second was the magnetic pull of past and future into an expanding conception of the present. In such a scenario, past and future are drawn into the present to be interpreted synchronically. Metaphorically, the timeline of history implodes onto a spatial plane defined by its topical places. In Hartog's scenario, contemporary history looked less like a temporal pattern, more like a spatial map of memory. His argument about presentism, therefore, suggested a direct connection to the memory phenomenon. Presentist history mimics many of the protocols of memory,

---

[38] For the range of historical approaches that might be identified as "postmodern," see the anthology edited by Keith Jenkins, *The Postmodern History Reader* (London: Routledge, 1997).

[39] François Hartog, *Regimes of Historicity; Presentism and Experiences of Time* (2003; New York: Columbia University Press, 2015), 8–11, 15–19.

substituting a mnemonic for a chronological framework of historical time. Its hallmarks include the present as the moment that defines the meaning of the past; the reading of history as a genealogical descent from the present; the instability of historical interpretation in the absence of a guiding master narrative; random time travel in search of a past relevant to present preoccupations; the abandonment of sequential narrative in favor of narratives departing from particular topics of historical interest.[40] In some ways Hartog's theory is reminiscent of the old and much maligned "philosophy of history" as a formula for grasping historical epochs since antiquity in their broad outlines. One might not be able to fix the past in an overarching pattern of events. But one can plot historical periodization in terms of changing conceptions of historical time by historians over the past 2000 years. Such conceptions shaped appreciation of the meaning of the past for the present.

All of these approaches—the end of history, postmodernism, presentism—reflect the nostalgic tenor of our times. Historian and public intellectual Tony Judt conveyed the sentiment well in his valedictory essay, *Ill Fares the Land* (2010), his bequest to a younger generation for rethinking the needs of the twenty-first century as they take up its tasks.[41] Judt interpreted the period 1955 to 1975 as a golden age from which statesmen have since beaten a retreat, with devastating consequences for permitting a widening divide between rich and poor, environmental degradation, and the corruption of liberal democracy by big money. The Left, Judt argued, had lost the momentum of its postwar energy for reform. Commitment to the welfare state, the great progressive project of the postwar era in Western Europe and America, was eroding; political neo-conservatism replacing it through an appeal to the populist cultural values of a bygone age and a retreat toward an economic doctrine of laissez-faire that threatened to undo hard-won social policies identified with the making of the welfare state.[42] While I am not suggesting that any of the historians of memory held these political views, many were attuned to Judt's pessimistic assessment of the dilemmas of the contemporary age.

As the idea of progress foundered, so the emotion of nostalgia became more visible in historical interpretations that scanned the past for missed opportunities that might have renewed the human prospect. In the midst of a time of uncertainty, when the near past disappointed and the future

---

[40] Ibid, 101–48.
[41] Tony Judt, *Ill Fares the Land* (New York: Penguin, 2010), 237.
[42] Ibid, 81–119.

promised diminished expectations, historians looked back into their own heritage for places of memory to which they could secure the historical moorings of the present age. The past was of interest not as seedbed for a unified narrative of human progress but rather for the fathomless pluralism from which lost worlds might be drawn.[43] As the German essayist Walter Benjamin remarked in an aphorism famous for its visionary nostalgia, we might look to neglected places of memory as if they were heliotropes turning "toward the sun that is rising in the sky of history."[44] That may explain why Benjamin has exercised such a magnetic influence upon historians of memory today in their efforts to identify missed opportunities on the path between past and present. Essayist Eva Hoffmann has characterized our time as an "era of memory."[45] Having lost a workable metanarrative, ours is a time for reflection on the way memory serves as a resource for finding a way out of our predicament. Among all the historiographical ventures of our times that have aspired to make historical sense of the present age, the discourse about memory was the one that would stick and generate new waves of scholarship that have rolled on for more than 30 years.

While attention to the topic of memory from the late 1970s led to labeling this historiographical phenomenon a "boom" in scholarship, some students of memory have pointed out that the topic of collective memory was not a completely new scholarly interest. Surveying the field, media scholars Astrid Erll and Ansgar Nünning argue that serious research on the topic dates from the late nineteenth century, while philosophical reflection harks back to antiquity. They propose that historical writing in the late twentieth century might better be characterized as a "second wave" of interest in the field.[46] The first wave was scientific; the second humanistic. The first had focused on the memory of the individual; the second on that of the social group. Therefore they argue for scholarly continuity between the two. The interest in memory has been ongoing,

---

[43] David Gross, *Lost Time; On Remembering and Forgetting in Late Modern Culture* (Amherst: University of Massachusetts Press, 2000), 22–24.

[44] Walter Benjamin, "On the Concept of History" (1940), in *Walter Benjamin; Selected Writings*, ed. Howard Eiland and Michael Jennings (Cambridge: Harvard University Press, 2003), 4: 390.

[45] Eva Hoffmann, *After Such Knowledge; Memory, History, and the Legacy of the Holocaust* (New York: Public Affairs, 2004), 241–44.

[46] Astrid Erll, "Cultural Memory Studies: An Introduction," in *Cultural Memory Studies; An International and Interdisciplinary Handbook*, ed. Astrid Erll and Ansgar Nünning (Berlin: Walter de Gruyter, 2008), 7–8.

they contend, gradually assuming larger proportions within diversifying scholarship not just in history but across the curriculum. Advances in scientific understanding of brain and memory correlate with those in the social sciences and the humanities.[47] Sociologists Jeffrey Olick, Vered Vinitzky-Seroussi, and Daniel Levy in their introduction to the *Collective Memory Reader* point to the sustained interest in collective memory by scholars in the social sciences over the course of the twentieth century. It is misleading, they argue, to contend that the seminal studies on collective memory by the sociologist Maurice Halbwachs during the interwar years had been largely forgotten by mid-century, only to be resurrected by Pierre Nora toward century's end.[48] If some historians thought of Halbwachs as a rediscovery, they contend, his work on collective memory had been a mainstay of sociological thought throughout the middle decades of the century. Historians of Halbwachs's day, moreover, built upon his legacy. As a professor at the University of Strasbourg during the 1920s, Halbwachs had worked in close association with pioneering scholars of the Annales movement in French historiography. Marc Bloch, they note, had written a thoughtful analytical review of his first book on the frameworks of social memory and commented on its correlation with his own research. While the topic of memory was somewhat neglected by the Annalistes during the era of the economic depression and World War II, they returned after the war to issues in collective mentalities that he and his colleague Lucien Febvre had pioneered in the early days of the Annales movement. During the 1960s, the history of collective mentalities was to become the leading edge of Annales scholarship in the making of a new cultural history. Their interest in collective memory appeared in the guise of tradition. Mentalities laid emphasis on collective memory in its habitual expression—customs, mores, common sense, rituals and festivals, folklore and folkways.[49]

---

[47] Some scientists have endeavored to bridge the gap between psychology, neuroscience and the humanities, notably Daniel Schacter, *Searching for Memory; the Brain, the Mind, and the Past* (New York: Basic Books, 1996); and Eric R. Kandel, *In Search of Memory; The Emergence of a New Science of Mind* (New York: Norton, 2006).

[48] Jeffrey Olick, Vered Vinitzky-Seroussi, and Daniel Levy, "Introduction," *The Collective Memory Reader* (Oxford, UK: Oxford University Press, 2011), 21–23.

[49] Philippe Ariès, "L'Histoire des Mentalités," in *La Nouvelle Histoire*, ed. Jacques Le Goff (Paris: Editions Complexe, 1988), 167–90.

FROM MENTALITIES TO MEMORY IN THE STUDY
OF CULTURAL HISTORY

Historian Alon Confino was among the first to note connections between scholarly work on collective mentalities during the 1960s and the turn to collective memory by the 1980s.[50] Mentalities, he explained, was a broad rubric, hazy in its definition, loosely conceived in its narratives, held together by a desire to address a variety of social attitudes that adhered stubbornly to the traditional way of life that held sway over European popular culture through the seventeenth century. As it emerged as an interest among the Annalistes, the study of mentalities was conducted within a structuralist paradigm, showcasing the inertial power of the past. Collective mentalities were mindsets steeped in repetition. The key to understanding mentalities was to grasp the weight of the past as carried forward in traditions that accommodated change only slowly and begrudgingly. Its long-range patterns were readily discernible. Large-scale anonymous processes dominated the historians' attention; individuals faded into the anonymity of group identity. Hence the history of mentalities betrayed a psychological determinism that acknowledged radical change only in rare moments of "conjuncture"—episodes of crisis when pent-up social and economic forces disrupted the equilibrium of deep structural stability.[51] Mentalities for historians, therefore, had been an exploration of a traditional way of life that had long since disappeared from the modern age. It was of interest now for the contrasts it revealed—a time when the inertial power of the past was a determining force in history, so different from our own in which forces of innovation convey an impression of incessant change and hence the illusion of the acceleration of time. By the 1960s, modern Western society was sufficiently different from those times for historians to look back upon them as a counterpoint to their own, sometimes with nostalgia as a "world that we have lost." It is no accident that the figure who emerged most prominently in scholarship on mentalities was Philippe Ariès, who in his writings sought to historicize beloved social and cultural traditions of his own heritage that had been swept away in the onslaught

---

[50] Alon Confino, "Memory and the History of Mentalities," in *Cultural Memory Studies*, ed. Erll and Nünning, 77–84; idem, "Collective Memory and Cultural History: Problems of Method," *American Historical Review* 105 (December 1997): 1388–90.

[51] Robert Mandrou, *Introduction à la France moderne, 1500–1640* (Paris: Albin Michel, 1974), 321–50.

of modernity from the time of World War II.[52] It is worth noting that Ariès's multi-volume *Histoire de la vie privée* was published in exactly the same years as Pierre Nora's *Les Lieux de mémoire*—1984–1992. The former summed up a generation of scholarship on mentalities; the latter set an agenda for the one coming of age about memory.

Mentalities may have been a springboard for the study of memory. Still, one might regard the interest in collective memory more as an offshoot from, than as a refinement of, an approach to mentalities, notably in the shift of focus from commonplace habits of mind embedded in immemorial tradition toward mnemonic practices that showcase memorable cultural representations. Put succinctly, mentalities was concerned with the conventional; memory with the exceptional. Mentalities focused upon the attitudes of ordinary people toward everyday life; memory studies, by contrast, gravitated toward the achievements of high culture, the memorable past created by artists, architects, and writers, as valued by critics and their educated audiences. Studies of collective memory turned toward cultural images and artifacts that appeal again and again across time as modes for remembrance of singular deeds (both good and ill), personalities, and aesthetic and intellectual accomplishments. Closely allied were biographies of personalities of exceptional and enduring interest—military, political, literary, and intellectual heroes. Examples include the American founding father, Abraham Lincoln, or reaching way back, Alexander the Great and Cleopatra.[53] Once they had been revered as life histories that inspired awe. Now historians redirected attention from their lives to their afterlives as memory figures, of interest for the way in which they were remembered by posterity as icons deployed and redesigned from time to time to suit the temper of the age.[54]

In contrast with collective mentalities' accent on stability, therefore, collective memory was of interest for its protean instability, dependent upon contingencies more than long-term patterns of repetition. Memory's episodes could intrude inconveniently into the present, jarring long-held

[52] Philippe Ariès, *Le Temps de l'histoire* (1953; Paris: Seuil, 1986), 222–23.

[53] For the hero as mnemonic icon, see, for example, Barry Schwartz, *Abraham Lincoln; Forge of National Memory* (Chicago: University of Chicago Press, 2000), 5–8, 293–312; Waldemar Heckel and Lawrence A. Tritle, eds., *Alexander the Great: A New History* (Chichester, UK: Wiley-Blackwell, 2009), esp. 218–310; Francesca T. Royster, *Becoming Cleopatra; the Shifting Image of an Icon* (New York: Palgrave Macmillan, 2003).

[54] Marek Tamm, ed., *The Afterlife of Events: Perspectives on Mnemohistory* (London: Palgrave, 2015).

habits of mind. The historiographical concept of collective mentalities had concentrated on what might be regarded as but one domain of collective memory. Reaching beyond, scholars inquired into strategies employed across the ages for preserving documents and cultural artifacts deemed worthy of remembrance in archives and museums. During the 1980s, the study of memory came into its own, transcending the boundaries of mentalities considered as its matrix. If the historians' interest in memory could not be understood as a totally new field of research in cultural history, it did pose a set of questions that would lead scholarly discussion in new directions. From that perspective, the memory phenomenon did signal a new departure in the historiography of the late twentieth century. To move from mentalities to memory was to witness a shift from received tacit understandings to consciously constructed mnemonic practices, understood in their varied possibilities. The idea of tradition was being reconceived as heritage, suggesting why the historic preservation move-ment gained new attention at about the same time.[55]

DEEP SOURCES OF THE LATE TWENTIETH-CENTURY MEMORY PHENOMENON

The historians' interest in memory from a critical perspective prompted them to search for antecedent scholarship earlier in the twentieth century. They settled upon three landmark scholars, memory icons in the study of memory: the sociologist Maurice Halbwachs on commemoration, the psychologist Sigmund Freud on trauma; and the student of language Walter Ong on technologies of communication. They would serve as prototypical models for memory studies within the social sciences and humanities in the late twentieth century. Each subtends a different field of memory studies analyzed in my following chapters: Halbwachs for his explanation of the way social power shapes the frameworks of collective memory; Freud for his clinical theory of the way the unconscious memory of trauma distorts conscious historical representation; Ong for his over-view of the way new technologies of communication have reshaped the uses of memory through the ages. Here is a synopsis of the significance of each.

---

[55] See David Lowenthal, *Possessed by the Past; the Heritage Crusade and the Spoils of History* (New York: The Free Press, 1996), 105–72, who draws distinctions between history and heritage.

Maurice Halbwachs (1877–1945)

The historians' late twentieth-century interest in the dynamic, socially conditioned nature of collective memory led to their reacquaintance with the seminal studies by the French sociologist Maurice Halbwachs during the 1920s.[56] Halbwachs argued that all personal memories are localized within social contexts that frame the way they are recalled. Without such social support, they tend to fade, for the way individuals remember is a function of the relative power of the social groups that influence them. Moreover, the particularities of personal memories in often repeated behavior are eventually worn down into social stereotypes. Only their most salient features stand out as these remembered episodes are telescoped into present consciousness. The long-range effect is to transform complex mnemic images into simplified eidetic icons. In this sense, collective memory is only residually the recollection of actual experiences, as its images are reconfigured to conform to contemporary cultural conceptions. In the process, the past is evoked in idealized expressions. Halbwachs tested his thesis in a case study of the localization of an imaginary landscape of the Holy Land by European pilgrims visiting Palestine from the fourth to the fifteenth centuries. Redeployed in the historical scholarship of the late twentieth century, his model came to serve as a prototype for method in this field.[57]

Sigmund Freud (1856–1939)

Halbwachs formulated his theory of collective memory as a refutation of the ideas of Sigmund Freud, whose best work focused on the psychology of the individual. Freud believed that individual memories remain intact in the recesses of the unconscious mind, from which they may be recovered through psychoanalytic technique. His method involved "working through" idealized "screen" memories that block access to the realities of traumatic experiences that induce forgetfulness. Recovery of these

[56] Maurice Halbwachs, *Les Cadres sociaux de la mémoire* (1925; New York : Arno Press, 1975), and the collection of his work: *Maurice Halbwachs, On Collective Memory,* ed. Lewis A. Coser (Chicago: University of Chicago Press, 1992).

[57] Maurice Halbwachs, *La Topographie légendaire des évangiles en Terre Sainte* (1941; Paris: Presses Universitaires de France, 1971). On Halbwachs as historian of memory, see Patrick Hutton, *History as an Art of Memory* (Hanover, NH: University Press of New England, 1993), 73–90.

repressed memories is the surest route to self-knowledge. In his later years, he expanded his theory to encompass collective memory, though scholars debate whether he explained adequately how such imagery is transmitted over time. His findings about the resurfacing of repressed collective memory, of the sort he presents in his *Totem and Taboo* (1913) and *Moses and Monotheism* (1939), flirt with the notion of a collective unconscious, one that today seems naive in light of all that we have since learned about cultural communication. The Freudian-inspired field of psychohistory, briefly prominent in the scholarship of the 1960s, eventually slipped to the margins of academic interest.[58]

While Halbwachs's claims for the social foundations of all memory may be exaggerated, his focus on memory's dynamic character was better attuned to late twentieth-century worries about memory's unstable nature. Still, the appeal of Freud's approach has persisted, particularly in the examination of unrequited trauma of the victims of the Holocaust. Freud's theory of repressed memory found new applications in scholarly efforts to put the legacies of these troubling memories in psychoanalytic perspective, both for surviving victims and for postwar statesmen in Germany seeking to come to terms with unresolved issues of guilt and responsibility for the genocide of European Jews. Among historians, their consideration was postponed for decades. The 1980s would be a time of sober assessment of the historical meaning of the Holocaust, notably in the debates of the German "Historians' Dispute" about whether its history could be integrated into a larger historical narrative so long as traumatic memory blocked access to the reality of what that experience had been for its victims. The issue of trauma demanded a different way of understanding memory in history. Screen memories masking traumatic experience served as place markers for opaque gaps in the historical record, signaling the disabling effect of repressed memories upon historical interpretation. New circumstances reminiscent of the old ones sometimes brought these latent anxieties out of hiding, allowing trauma to fester openly.[59] But it would take time and the work of psychoanalysis to lift the hidden meaning of these unconscious memories into conscious understanding. Freud's theory of

---

[58] An exception is Peter Gay's skillful integration of Freudian insight into his work on European cultural history. See esp. his *Freud for Historians* (Oxford: Oxford University Press, 1985).

[59] Henry Rousso, *Le Syndrome de Vichy de 1944 à nos jours* (Paris: Seuil, 1987), 13–20, 118–54.

repression prompted historians of the Holocaust to take a hard look at these long-term effects of trauma on postwar politics, and beyond into our own times. Coming to terms with the historical significance of the Holocaust became a major scholarly preoccupation during the 1980s in Germany and in France, and Freud's theory was initially invoked to deal with it. The panorama of late twentieth-century memory studies, therefore, may be read as a tension between the prototypical models devised by Halbwachs and Freud.[60]

Walter Ong (1912–2003)

Ong is the key theorist for a third pathway into memory studies that emerged independently during the early twentieth century among students of ancient oral tradition. Inspired by the revolution in technology of electronic communication, Ong is noteworthy for taking the longer view of this approach to collective memory. He recognized the significance of corresponding transitions in the past, notably in Greco-Roman antiquity in its passage out of primary orality into manuscript literacy, and again in the democratization of print culture during the European Enlightenment. Ong was a student of Marshall McLuhan, famous for his celebration of new media, and his interpretation might be construed as derivative of his mentor's insights. Ong's *Orality and Literacy* (1982) provides a lucid synthesis of a generation of scholarship on these transitions for transforming the uses of memory and so became the standard text for understanding its history.[61] A change of technology precipitates a change of mindset, he argued, which includes changing modes of perception, learning, and the organization of knowledge.

Ong's interpretation is foundational for promoting a number of propositions about collective memory in cultural transmission: No technology of communication ever disappears. Rather, it is nested within the new medium, complicating and enriching cultural interchange. Literacy incites an aspiration to stabilize the memory of the past, and so creates preconditions for the preservation of knowledge over long periods of time, a topic German scholars Jan and Aleida Assmann would address in depth

[60] Patrick Hutton, "Sigmund Freud and Maurice Halbwachs: The Problem of Memory in Historical Psychology," *The History Teacher* 27 (1994), 145–58.

[61] Walter Ong, *Orality and Literacy; The Technologizing of the Word* (London: Methuen, 1982), 78–116.

under the rubric "cultural memory." The transition from primary orality to manuscript literacy in Greek antiquity sparked the debate set forth deep in antiquity by Plato in his Socratic dialogue *Phaedrus* (circa 370 BCE) about the respective merits of each as an art of memory. Ong wrote with less authority about the transition to electronic communication in our own times, but offered the intriguing suggestion that we stand at the threshold of a "secondary orality."[62] This notion would be developed by students of communication from the turn of the twenty-first century under the rubric "remediation." Their emphasis upon the dynamic, ceaselessly changing nature of memory as communicated by media resonates with the mnemonic operations of primary orality in unleashing memory from the binding forms of print literacy, permitting it to roam freely in cyberspace in unprecedented ways. Orality/literacy as a topic for historians has been comprehensively studied, and by the turn of the twenty-first century no longer served as the cutting edge of this line of historiographical inquiry.[63] Ong is, nonetheless, a deep ancestor for today's work on the cultural effects of the technologies of digital age communication.

Aby Warburg (1866–1929)

I should say a few words, too, about the rehabilitation of the ideas of the German art collector Aby Warburg. He would be recalled into the discussion of cultural memory by Jan and Aleida Assmann as a forerunner of the idea of memory's time traveling, a topic of gathering interest in our digital age. He was not a prolific author and what little he wrote about cultural memory is recondite.[64] He came of age in postwar Germany. Independently wealthy and well-to-do, he was a connoisseur of art and artifacts of the high culture of the Renaissance era. He wrote for arcane publications of interest largely to antiquarians. But his accomplishment as a collector was extraordinary. Warburg died in 1929, and in 1936 his heirs moved his vast collection of memorabilia to the University of London. There the collection would become the core of the holdings of the institute that bears his name, best known to students of memory

---

[62] Ibid, 135–36.

[63] See also the overview by the anthropologist and historians of African oral tradition, Jan Vansina, *Oral Tradition as History* (Madison: University of Wisconsin Press), 1985.

[64] For Warburg's theory of social memory, see Ernst Hans Gombrich, *Aby Warburg; An Intellectual Biography* (Chicago: University of Chicago Press, 1986), 239–59; Alon Confino, "Collective Memory and Cultural History," 1390–92.

for the scholarship that Frances Yates was able to draw from its archives in the years after World War II.[65] Warburg believed that figural representations of archetypal human experience are capable of triggering emotions that resonate with experience in like circumstances throughout the ages. He was especially interested in pictorial art that displays mnemic images of universal human appeal, deep traces of what we used to call "classical" poses significant for their capacity to evoke timeless human emotions. Warburg studied those created in Greco-Roman antiquity that inspired and hence were incorporated into the art of the Renaissance.[66] An image is not simply rediscovered, but is reused in new contexts.

Warburg's work does not stand alone as an approach to the time travel of memory. It has affinities with Walter Benjamin's conception of the "profane illumination" of experience transported out of an earlier age to enliven creative insight in the present.[67] Warburg's idea of time travel also invites comparison with Arthur Lovejoy's conception of the history of ideas, presented as a new field of historiography during the 1930s. For Lovejoy, a seminal idea would pop up over the course of time in new contexts and a variety of ways.[68] So evoked, the idea is not only recalled but recirculated and recast in new settings. Warburg's approach, however, was more encompassing, for he sought to combine images with ideas, taking into consideration the basic human emotions they incite in any age. Ideas, he contended, are not easily extricated from the mnemic images in which they are conveyed, nor are the emotions they arouse. Warburg's work, therefore, has taken on even more interest among today's students of media as a prefiguration of the kind of mnemonic time travel showcased in electronic media in all of its varieties—uplifting from the past photos, film, video animation, television clips, as well as the older technologies of print culture into cyberspace for recycling in the digital modes of

[65] Frances Yates, "Autobiographical Fragments", in *Collected Essays*, ed. J. N. Hillgarth and J. B. Trapp (London: Routledge and Kegan Paul, 1984), 3: 316–19.

[66] Aby Warburg, *The Renewal of Pagan Antiquity*, trans. David Britt (Los Angeles: Getty Research Institute, 1999); see also Fritz Saxl, "The History of Warburg's Library," in Gombrich, *Aby Warburg; An Intellectual Biography*, 325–38.

[67] Walter Benjamin, "Excavation and Memory" (1932), in *Walter Benjamin; Selected Writings*, ed. Michael Jennings et al. (Cambridge: Harvard University Press, 1999), 2: 576.

[68] See Lovejoy's essay "Reflections on the History of Ideas," together with his exchange with critics in *The History of Ideas; Canon and Variations*, ed. Donald R. Kelley (Rochester: University of Rochester Press, 1990), 1–70.

present-day consumption.[69] Here such images recombine in the synergy of these varied modes of representation beyond anything Warburg himself might have imagined. But the idea about the quickening effect of memory in the reuse of old images in new creative settings was originally his own.

PLAN FOR THE BOOK

I pursue two avenues of approach. I begin with a phenomenological description of the rise of historical scholarship on collective memory. I retrace three royal roads along which pioneering studies once traveled: the politics of commemoration (purposes and practices of remembrance); the cultural effects of historical transitions in the invention and uses of new technologies of communication (from primary orality to media culture via manuscript and print literacy); the effects of trauma upon memory (with particular attention to the memory of the Holocaust). These pathways were traveled independently during the 1960s and 1970s, though in time they would converge. The 1980s is the crucial decade in which connections among them became visible. By the turn of the twenty-first century, the work of historians would mix with that of scholars from across the humanities and social sciences to reemerge under the heading "memory studies," reconceived as an interdisciplinary venture.

In my second avenue of approach, taken up in later chapters, I analyze the historiographical implications of this scholarship for understanding how critical inquiry into the workings of memory has changed our thinking about history itself. I pay particular attention to chains of memory in their transfiguration, the mnemonics of historical periodization, the turn from progress to nostalgia as a perspective on historical time, the emergence of the present as the privileged moment of historical time in memory studies, and with it the revival of an interest in history as living performance as opposed to its representation at a critical remove. I offer some thoughts on the way we have come to understand how history and memory, as different as their resources may be, have found a middle ground in the concept of historical remembrance. As a context for their discussion, I develop two perspectives. The first perspective considers history from the vantage point of the mnemonics of time in light of the interpretations of such scholars as Reinhart Koselleck, Peter Fritzsche, Svetlana

---

[69] Aleida Assmann, *Cultural Memory and Western Civilization* (Cambridge: Cambridge University Press, 2011), 214–18, 220–21, 358–60.

Boym, and François Hartog on the varied ways in which historical time may be configured. The second concerns the historians' interest in memory as a reaction to the rhetorical turn in scholarship (noting the shifting scholarly appreciation of Michel Foucault from his method as historian to his life experience as an anecdotal example of this redirection). Here I address the tension between memory as representation and as experience, as explained by Frank Ankersmit and Jay Winter, with some attention to the performative modes of historical remembrance (as in drama, tourism, and historical reenactment). I close with a discussion of historians who reflect on history's role in our present-minded digital age, among them Gavriel Rosenfeld, Arlette Farge, Robert Darnton, Yosef Yerushalmi, and Paul Ricoeur.

Memory studies, like all historiographical fashions that preceded it, have over time diversified and expanded their range of inquiry. Of particular interest for historiography, scholars have formulated ever more sophisticated interpretations of the nature and possibilities of collective memory. Today there is more attention to the way memory is embedded in a wide range of social and cultural practices. As historian Jay Winter has pointed out, scholars explain the way memory as a faculty of mind enables humans to fight back against the forgetfulness that postmodern consumerism promotes, and so inspires not mere remembrance but creative engagement with the past as a resource for envisioning the future.[70] The coming of the memory topic to prominence in our times was first heralded by the publication of a monumental project, designed and orchestrated by French historian Pierre Nora, the three volume *Les Lieux de mémoire* (1984–1992). This project would set the tone for decades to come. French in origin and orientation, its influence would be felt around the world. My following attention to the work of German scholars Jan and Aleida Assmann on the elaboration of cultural memory serves as a counterpoint to Nora in their focus on the elaboration of cultural memory from antiquity into modern times. Herein history plays only an ancillary role as one among the many arts of memory. It is to these beginnings of the memory phenomenon that I now turn.

---

[70] Jay Winter, *Remembering War; The Great War Between Memory and History in the Twentieth Century* (New Haven, CT: Yale University Press, 2006), 288–89.

CHAPTER 2

Pierre Nora's *Les Lieux de mémoire* 30 Years After

LES LIEUX DE MÉMOIRE AND THE CRISIS OF CONTEMPORARY FRENCH IDENTITY

French historians have long aspired to place themselves within the avant-garde of historical writing. From Jules Michelet to Fernand Braudel, French historiography has enjoyed international prestige for pioneering new directions in historical research. Pierre Nora (b. 1931) has furthered this venture by taking French scholarship into an unexplored realm of intellectual inquiry in the late twentieth century, thanks to his project on the deconstruction of the French national memory. His *Les Lieux de mémoire* (1984–1992) serves as a landmark in the emergence of memory studies, not only in France but around the world.[1] Nora presented his project as a collaborative enterprise of some 125 well-known colleagues. But its overall conception was very much his own. His explanatory essays frame its organization and guide its reading. Though he focused exclusively on France, *Les Lieux* was recognized at once as a foundational study in the politics of commemoration, and subsequently as a turning point in the conceptualization of historical writing for our times. The success of its reception surprised everyone, perhaps no one more than Nora himself.

Nora arrived on the scholarly scene at an opportune moment to pursue this project, a time in which the leading traditions of French historical

---

[1] (Paris: Gallimard, 1984–1992), 3 vols.

© The Editor(s) (if applicable) and The Author(s) 2016
P.H. Hutton, *The Memory Phenomenon in Contemporary Historical Writing*, DOI 10.1057/978-1-137-49466-5_2

writing were visibly losing the force of their once considerable cachet. Marxism no longer exercised the mystique it had once held for left-wing intellectuals of the postwar era. The Annales movement, with which cutting-edge research in social and economic history had been identified since the mid-twentieth century, had shed the evangelical fervor of its beginnings in the interwar years. The grand ambition of its founders, Lucien Febvre and Marc Bloch, to write a total history had stalled, as these pioneers of the Annales movement were followed by settlers content to articulate the paradigm. A sure sign of the way innovation had yielded place to convention was the elevation of its founders as commemorative totems of its beginnings.[2] A younger generation of Annales scholars, Nora among them, codified its accomplishments in encyclopedias and handbooks during the 1970s.[3]

Significant in a more general way was the waning influence of the French revolutionary tradition as the foundational frame of reference for writing about French history. That tradition had been sustained by a variety of ideologies born of the matrix of the French Revolution—liberalism, radicalism, communism, and nationalism the most prominent among them, not to mention a resurgent royalism as their persistent adversary. Those who wrote about modern French history tended to identify with one or another among them, their fairness and objectivity notwithstanding. By the 1970s, however, historical writing inspired by ideological conviction— whether of the Left or the Right—was losing its appeal for French historians. French historical scholarship was in the process of cutting itself free from its deep roots in the revolutionary tradition. With that separation, the broadly conceived narrative of the rise of the modern French nation-state that had sprung from that tradition lost its coherence. The decline of the conception of French history as an ongoing narrative driven by expectations of the future signified a broader historiographical trend. Scholarship across the humanities and the social sciences was taking a rhetorical turn in which scholars sought to deconstruct narratives to understand the bias of their composition.[4] Nora's project was conceived within this context.

---

[2] Patrick Hutton, "France at the End of History; The Politics of Culture in Contemporary French Historiography," *Historical Reflections* 23/2 (Spring 1997), 105–27.

[3] Jacques Le Goff, ed., *La Nouvelle Histoire* (Paris: Editions Complexe, 1978); Jacques Le Goff and Pierre Nora, eds., *Faire de l'histoire* (Paris: Gallimard, 1974), 3 vols.

[4] Hayden White, *Metahistory; The Historical Imagination in Nineteenth-Century Europe* (Baltimore: Johns Hopkins University Press, 1973); David Harlan, "*Intellectual History and the Return of Literature*," *American Historical Review* 94 (1989): 581–609.

He aspired to inventory the many places of memory—some marginalized or obscured by the force of the revolutionary tradition—that had contributed to conceptions of French identity along the way.[5]

By the time that Nora embarked on this venture during the late 1970s, France was no longer the nation once celebrated in the grand narrative about its making. Its stature as a nation-state had diminished over the course of the twentieth century. France had accepted a humiliating armistice with Nazi Germany at the outset of World War II, and the complicity of the Vichy regime in the Holocaust took decades to work through in critical historical examination. Belated cases against collaborators taken up during the 1980s were a cause of public embarrassment. France, moreover, had been obliged to surrender its worldwide colonial empire, in the case of Algeria with considerable strife. France retained its stature as an important nation-state, but now as one among many in an emerging economic and political confederation of Europe. Even for its high culture, France no longer enjoyed unchallenged pride of place. Francophiles still delighted in its beautiful language. But as a lingua franca for science and commerce, French had been obliged to make way for American English in an age of globalization. Possibly most important, the dramatic popular upheavals of the nineteenth century that had lent credence to the idea of a revolutionary tradition were now fading memories.

Such were the issues that converged to raise questions about the nature of France's historical identity as it might be understood in the present age. What remained to bolster national pride was a heritage, a resource of great complexity whose story Nora wanted to tell in an innovative way. He envisioned a new paradigm for writing French history, one that was less an interpretation of its realities, more a reflection on the imaginative ways in which its identity had been represented through the ages. Like the founders of the Annales movement who had set an agenda for historical research in the early twentieth century, Nora set another at century's end. It was to be no simple task. He notes that he devoted a decade of scholarship to its conceptualization. His reflections on the historiographical implications of his project continue to this day.

As Nora addressed the editorial tasks of this venture, the bicentenary of the French Revolution loomed on the horizon. The memory of the Revolution, dramatized in images of its popular insurrections, had been the foundation of the French national identity. Drawing on the political

---

[5] Pierre Nora, "Entre mémoire et histoire," in *Lieux de mémoire*, 1: xviii–xli.

tradition that the Revolution had inspired, France had provided leadership among the European nations in forging democratic institutions within the framework of a liberal society.[6] By the late nineteenth century, the construction of a republic on stable institutional foundations was celebrated as a hard-won accomplishment, not only for the triumph of its principles over rival models of government, but also for its role in promoting science and the humanities, hallmarks in the making of a modern way of life.

But on the eve of the twenty-first century, Nora asked, was the legacy of the French Revolution any longer an adequate frame of reference with which to evaluate the newly emerging realities of French identity in the late twentieth century? The once powerful image of France as a nation of small property owners in rural villages and small towns, politically democratic yet socially conservative, had become an obsolete cliché. France had become more urban, more heterogeneous in its population, more aware of its regional diversity, more caught up in the globalizing economy of a consumerist culture. The French had celebrated the centennial of the Revolution with a certain harmony and satisfaction. Preparations for the bicentennial, by contrast, were fraught with controversy. There was little consensus about how it should be celebrated.[7] One well-known scholar of the Revolution facetiously questioned whether it should be celebrated at all.[8] Certainly not in the way it had been 100 years before. If the Revolution was no longer the matrix of the national identity, how then should that identity be reconceived for the present age? Was it not time, Nora wondered, to inquire once more into deep sources of the French national heritage, evoking not only well-known commemorative images of recent origin but also others that were residues of once bright and vital memories, now long lost and forgotten.

---

[6] David Thomson, *Democracy in France since 1870*, 5th ed. (Oxford: Oxford University Press, 1969); François Furet, *Lies, Passions and Illusions*, ed. Christophe Prochasson (Chicago: University of Chicago Press, 2014), 75–81.

[7] Steven Laurence Kaplan, *Farewell Revolution; the Historians' Feud, France 1789–1989* (Ithaca: Cornell University Press, 1995).

[8] François Furet, "Faut-il célébrer le bicentenaire de la Révolution française?" *L'Histoire* 52 (1983):71–77.

MEMORY PALACES FOR THE PLACES OF THE FRENCH NATIONAL MEMORY

Nora was ideally suited to undertake this project. As a young scholar, he had written two important articles on Ernest Lavisse (1842–1922), a magisterial figure in the professionalization of historical writing in the late nineteenth century, and, one might say, a leading proponent of the grand narrative of French history. Lavisse had worked in the positivist tradition of Auguste Comte in an effort to move historical scholarship out of the realm of literature into that of social science. As a professor at the Sorbonne, he had presided over an entourage of young scholars who enthusiastically followed him into the archives for endless hours of research. Lavisse was also a leading figure in educational reform. His multi-volume history of France became a primer for French public schools. Well connected with leading statesmen of the turn of the twentieth century, he had played a key role in providing an intellectual apology for the liberal values of the Third French Republic. Nora's study of Lavisse not only enabled him to understand the bias of historical writing in that era but also gave him a frame of reference for assessing the historical perspective of his own. He noted that Lavisse, for all his research, scientific rigor, and sense of civic purpose, wrote a history that was profoundly grounded in an unacknowledged memory of the origins and development of France as a nation-state. Lavisse presented the Republic as the instrument of the civilizing process, and as such endowed with high moral purpose. Such a history emphasized the continuity of the story of France from its medieval beginnings. It presumed a sense of direction.[9] Nora, by contrast, sought to explain why that conception of history had lost its meaning for the present age.

Nora launched his project modestly in the late 1970s as a seminar at the Ecole des Hautes Etudes en Sciences Sociales, the prestigious French graduate college to which he had recently been elected. He claims that this venture was initially experimental, an open-ended excursion into the sources of the French cultural heritage. He is quite specific about the time that collective memory as a concept for exploring that heritage became a topic of particular interest to historians. He pinpoints 1970–1980 as the

---

[9] Pierre Nora, "L'Histoire de France de Lavisse," in *Lieux de mémoire*, 2: 317–75; idem, "Pourquoi lire Lavisse aujourd'hui?" in *Présent, nation, mémoire*, ed. Pierre Nora (Paris: Gallimard, 2009), 193–204.

crucial decade.[10] In his public role, moreover, Nora doubled as an editor at the Gallimard publishing house. He had an insider's knowledge of the best current research and was able to call upon eminent scholars to write about particular places of memory that he wanted to include in his ambitious enterprise. The project grew as it moved from his seminar to his editorial offices. Its scope morphed from three into seven large tomes over the course of the 1980s.

As a point of departure, Nora took a lesson from the early twentieth-century French sociologist Maurice Halbwachs, who had written seminal studies about the workings of collective memory. For Halbwachs, collective memory provides the source material for history. But such memory is sustained by social power. The rise of professional history, Nora perceived, was coeval with the rising power of the bourgeoisie to fashion the nation-state in its own image. As its role expanded and its prestige grew over the course of the nineteenth century, the writing of history took its cue from this sustaining memory of its civilizing role. Nora further reasoned that as the nation-state as a magnet of social allegiance weakened in the late twentieth century, so too did the grand narrative of French history on which it was based. It is worth noting that only a few years before the first volume of Nora's project on the national memory appeared, philosopher Jean-François Lyotard famously wrote a widely read book about the postmodern abandonment of the grand narrative of history.[11] That dissolution of the story of modern French history became the backdrop of Nora's research project. As the capacious collective memory of the rise of the nation-state lost its force, the past was opened to a variety of alternatives. Each one offered a perspective on a particular aspect of French identity. A coherent story gave way to the aggregation of many.

Here Nora borrowed the idea of places of memory from a book on the ancient art of memory by the English historian Frances Yates. For her, the art was more than a method of memory retrieval, or even an appreciation of the ornate "memory palaces" that Renaissance cosmologists constructed on its principles. Yates's method was philological. Places of memory were points of intellectual departure. One returned to such places to follow the stories they had inspired. The chains of memory they narrated were over time revised, and their meanings assumed different

---

[10] Pierre Nora, "Les Trois Pôles de la conscience historique contemporaine," in *Présent, nation, mémoire*, 13.

[11] Jean-François Lyotard, *La Condition postmoderne* (Paris: Editions de Minuit, 1979).

forms.[12] Places of memory, therefore, were references for writing a history that reconstructed the past as it had been imagined—as places on a map of memory. Nora's plan for exploring the French national memory, therefore, borrowed her idea of spatial design. He framed his study as a repertoire of maps, each one grouping related memories in the conceptualization of the French national identity—republic, nation, cultural heritage. As an art of memory, his scheme could be read in two ways: genealogically and culturally. It was to be read genealogically as a descent from the present into the past. Culturally, it was read as a move from concrete toward abstract notions of the French heritage.[13]

Still, this tripartite notion of Nora's design does not adequately convey the multitude of places of memory that populate these schemes, whose diversity he and his contributors set about to explore. His three principal repertoires of maps might be likened to memory palaces, but on a scale beyond any that Yates's Renaissance philosophers might have imagined. Within each palace, he identified conceptual networks that might be characterized as its hallways. His own essays—ten in all—opened doors to these passageways. Among the networks of the memory of the Republic, he included such imaginary concepts as symbols, pedagogy, commemorations, and counter-memories of its opponents. For the nation, he organized his scheme around mnemonic notions of histories, landscapes, monuments of the state, past glories, beloved writers. For his third palace, *Les France*, the coordinating sinews included such categories as conflicts over identity, traditions, ways of life, the archives, and emblems that housed profound secrets of the French cultural heritage. Linked by these networks, the articles of his contributors served as rooms in the palaces, each one anchoring a different place of memory. Nora included some 125 contributors, all French with the rare exception. *Les Lieux*, therefore, consisted of 128 articles: 18 for the Republic, 48 for the Nation, 62 for *Les France*.

Did such an aggregation of particular memories of France permit a unified conception of what it means to be French? Nora conceded that his repertoire of maps of French memory did not display the readily visible unity of the grand political narrative about the rise of the nation-state. But unity, he believed, could no longer be conceived exclusively

---

[12] Frances Yates, *The Art of Memory* (Chicago: University of Chicago Press, 1966).

[13] Patrick Hutton, *History as an Art of Memory* (Hanover, NH: University Press of New England, 1993), 147–53.

in political terms with well-defined territorial borders. Here he treated France as an imagined community bound together by the network of its places of memory. Each site reflected all of the others, he remarked in his essay prefacing his third volume: "Comment écrire l'histoire de France." As he explained: "Each of these essays is a profound sounding, a view of France from a fly's eye, a crystal ball, a symbolic fragment of a symbolic ensemble. There may well be a unified France, but none of these subjects, these objects, or these 'places' would serve as the foundation for a unified history of France. Each is all of France, according to its manner."[14]

NORA'S REFLECTIONS ON *LES LIEUX* 30 YEARS AFTER

As the memory phenomenon in contemporary scholarship took hold by the end of the twentieth century, Nora's stature as an historian grew. In 2002, he was elected to the Académie Française. In 2011, an admiring colleague wrote a highly sympathetic biography.[15] Nora was called upon to speak everywhere about *Les Lieux*, and he composed numerous essays along the way concerning the historiographical implications of his project—both its strategy for writing French history and its re-visioning of French historical identity for the contemporary age.[16] In a recent reflection on the significance of his project nearly 30 years after the appearance of the first volume of *Les Lieux*, Nora placed his accent on the context in which his project might best be understood in light of France's historiographical traditions.[17] In *Les Lieux*, he had organized his historical places of the French memory as a genealogical descent: from republic, to nation, to *Les France*. In this essay, he reverted from the spatial design of *Les Lieux* to a more conventional linear one that demarcated the ascending stages of the history of modern French historical writing. He outlined four broadly conceived stages in the changing historical consciousness of modern France, considering each in light of its understanding of memory. He dealt with the first two stages quickly. Early nineteenth-century

---

[14] Pierre Nora, "Comment écrire l'histoire de France," in *Lieux de mémoire*, 3: 22–23. The conceptual coherence of Nora's project was lost in translation as it was parceled out into two separate American editions. American publishers divided up and abridged the project, picking and choosing among its essays.

[15] François Dosse, *Pierre Nora; Homo Historicus* (Paris: Perrin, 2011).

[16] Pierre Nora, *Historien public* (Paris: Gallimard, 2011), 11–15; idem, "Les Trois Pôles de la conscience historique contemporaine," 26–29.

[17] Nora, "Les Trois Pôles de la conscience historique contemporaine," 7–29.

historiography was romantic in its desire to make the memory of the past live again. The writing of such history aspired to touch living memory directly, as noted especially in the writings of Jules Michelet. For this much beloved historian, modern French history was visibly inspired by the revolutionary tradition, as he evaluated the past in light of his personal witness of the revolution of 1830. In the late nineteenth century, by contrast, historiography had come to wear the mantle of social science in the manner prescribed by Lavisse. Living memory sustained the writing of history, but as a tacitly understood resource.

The heart of Nora's account, however, was his contrast between the historiography of the Annales school and that of his own mnemonic turn. For the better part of the twentieth century, Annales scholarship had claimed the title of the "new history." But for Nora, the new history of the Annales had grown old, whereas memory, long regarded as the unwritten source of history, had come forth from the shadows to serve as subject matter for historical scholarship in the present age. Annales scholars had been interested in collective memory, but primarily as the deep source of attitudes embodied in the collective mentalities of a rural, preindustrial society whose habits of mind lingered into the modern age even though its customs were dying away. Here Nora was able to contrast the history of collective mentalities, the last realm of Annales historiography during the 1960s, with his own history of collective memory to dramatize the shift in historical thinking that was under way. The Annales interest in barely perceptible change over long periods of early modern history (*histoire à la longue durée*) gave way to an appreciation of the accelerating pace of change in the present age. The Annales investigations of the workings of living tradition to illuminate long-term cultural continuities yielded place to Nora's inventory of the ever more frequent erection of commemorative sites, which he interpreted as efforts to resuscitate tradition's waning authority. The historical study of the past as a way to anticipate the future was abandoned in the knowledge that today we no longer sense that the past sustains us, any more than we possess a sure expectation of where history may be tending.

Nora also offered a new set of mnemonic places as a basis for this historiographical reflection. It is instructive to juxtapose the old set to the new. The historical places of *Les Lieux*—Republic, Nation, *Les France*—were here set aside in favor of historiographical ones: present, nation, memory. Each, Nora commented, has profound implications for what his project revealed about a change in historical consciousness in our times. As he put it somewhat dramatically, "the decade 1970–1980 is the one that has witnessed the most important mutation in the national memory (*modèle*) since the revolutionary decade of 1789."[18] He characterized these places of today's French historiography in the following three themes:

The Present Age has Shed Its Ties to the Past

Nora argued that the present age is not an extension of the past, but rather an age apart. Lavisse's conception of modern history, he remarked, had been future-oriented. It looked to origins and it anticipated a direction of progress through reform. Contemporary historical scholarship, by contrast, is present-minded. It has no foundational touchstone of the sort that Lavisse's *mémoire/histoire* ascribed to the French Revolution. Nor does it anticipate what might lie beyond the horizons of the future. It does not look to the near past for reassuring continuities but rather to random places of memory both near and far. Topics localized there are imported into the present insofar as they speak to present needs. The present age, therefore, is a time of memory—"an era of commemoration," as Nora put it.[19] In effect, he believed, the intense interest in memory in relation to history is symptomatic of a new way of thinking about historical time. Here he drew on the scholarship of François Hartog, who introduced the notion of "regimes of historicity" to characterize these shifts in the perception of historical consciousness.[20]

The Nation and the Waning of the French Revolutionary Tradition

As for the French nation, its changing politics reflected the changing identity of its culture. The notion of republican/royalist rivalry issuing from the combats of the nineteenth-century revolutionary tradition had been laid to rest. Its late twentieth-century avatars were the Communist Party and the faithful disciples of Charles de Gaulle. Both had been mainstays of the French resistance during World War II and so renewed the memory of the revolutionary tradition in the postwar era. By the 1970s, however,

---

[18] Ibid, 13.
[19] Pierre Nora, "L'Ere de la commémoration," in *Lieux de mémoire*, 3: 977–1012.
[20] François Hartog, *Régimes d'historicité; présentisme et expériences du temps* (Paris: Seuil, 2003).

the Communist Party had become a senescent, sectarian organization, conspicuous for its outdated doctrinal orthodoxy. Its stance on the student "revolution" of 1968 was noteworthy for its refusal to take seriously the grievances of this youth movement for educational and environmental reform. By the mid-1970s, the Party had succumbed to what Nora characterizes as the "Solzhenitsyn effect": the brutal record of its dictatorial forbearers had been too starkly exposed to merit apology any more. Communism, irreparably vitiated by its Bolshevik heritage, was dying. The fortunes of Gaullism played out somewhat differently. As a coherent political movement that embodied French nationalism, the Gaullist coalition appeared to have disintegrated upon the death of de Gaulle in 1970. But his memory lived on as his nationalism of statecraft was transfigured into patriotism associated with the larger French cultural heritage.[21]

This subtle transition from ideology to heritage was to be observed as well, Nora argued, in the emergence of the hybrid politics of presidents Valéry Giscard d'Estaing and François Mitterrand during the decade 1975–1985. Giscard, a later-day embodiment of Orléanist elegance, combined respect for tradition with support for technological initiatives. Mitterrand, the first professed "socialist" to win the presidency, was a humanist and a cultural Catholic. Their pragmatic policies, Nora contended, signaled the end of the old ideologies as defining forces in French politics. The revolutionary tradition had come to be identified with a dated way of thinking. As its influence faded, a middle-of-the road politics emerged to guide French public affairs.[22] About the same time, François Furet, then a leading historian of the French Revolution, emphasized the consolidation of governmental power and reduced ideology to the status of an imaginary discourse in his provocative reinterpretation of its long-term legacy.[23]

Memory Then and Now

Given the waning authority of the revolutionary tradition, Nora argued, collective memory ceased to sustain historical interpretation in the way it had a century before. For Lavisse and his colleagues, he noted, collective memory had provided centuries of staying power upon which they might

---

[21] Pierre Nora, "Gaullistes et Communistes, in *Lieux de mémoire*, 3: 347–393; idem, "L'Historien devant de Gaulle," in *Présent, nation, mémoire*, 278–88.

[22] Nora, "Les Trois pôles de la conscience historique contemporaine," 13–19.

[23] François Furet, *Penser la Révolution française* (Paris: Gallimard, 1978), 15–16, 46, 49.

draw to correct its distortions and remedy its deficiencies. At the same time, historians had cautioned one another not to stray too close to the present (roughly defined as the span of three generations), for they judged living memory unreliable and resistant to critical interpretation. For historians of that epoch, it took time for testimony to coalesce into usable evidence. But so much had the chain of the French national memory been fractured over the course of the twentieth century, first by war and then by fast-moving technological change, that yesterday's wisdom was today dismissed with hardly any regret. What counted for understanding change in the present age was rapid, ceaseless innovation. As the power of collective memory born of the experience of the past waned, that of living memory in the present ironically acquired ever greater intensity. Living memory, once identified with the wisdom of the ages, was reconceived as the experience of the present age. What had been perceived to be the deep living sources of the national history were reduced to lifeless remains, accessible only in their commemorative representations.

The effect was two-fold. First, a newfound interest in commemoration burst forth on the public scene during the 1970s and quickly became a national obsession.[24] Commemorative events took off in a way that exemplified the democratization of public memory. The new enthusiasm for memory was made manifest especially in the historic preservation movement, which vastly widened the sphere of its concerns. The French had long sought to refurbish their decaying cathedrals and chateaux. But now mementos of a vanishing rural way of life came to the fore. As the young escaped the small towns and villages of France, living memory of a way of life that had roots in the Middle Ages yielded to the commemoration of its passing. This expansion of commemorative practices signified nostalgia for a disappearing way of rural life in the face of an emerging urban economy and the growing ethnic diversity of the French population. The cause today among the thousands of emerging local historic preservation societies, Nora remarked, is "to save the village laundry and the cobblestones of the old streets."[25]

Second, and as a counter-current, living memory was reborn as the mode of contemporary history, stimulating popular interest as never before. Witnesses to the great or catastrophic events of our times offered

---

[24] Pierre Nora, "Reasons for the Current Upsurge in Memory," *Transit* no. 22 (4 April 2002): 1–8.

[25] Nora, "Les Trois Pôles de la conscience historique contemporaine," 18.

testimonies. Statesmen wrote accounts of their years in office. Even historians composed memoirs (*égo-histoires*) of their entry into and progress within the profession. On the popular level, the historical reenactment movement took off, drawing in cadres of history buffs who wished to experience vicariously "what life was like" in the events that they dramatized in costume and ritual performance. Living memory was perceived to be the real "existential" history, and as such was given fuller attention than ever before. But the span of such living memory had contracted narrowly around the generation that had come of age. In this way, living memory became a surrogate for contemporary history, for what mattered most was the present age conceived as the privileged moment of historical time.

NORA AND HIS CRITICS

Paul Ricoeur: "L'Insolite Lieux de mémoire"

The most searching critique of Nora's project was offered by philosopher Paul Ricoeur (1913–2005), a leading phenomenologist of his day. A decade after the final volume of Nora's project appeared, Ricoeur engaged him in a polite exchange. It was one between two very different kinds of scholars, and one might say personalities. Nora was the historiographer par excellence. He had pioneered a new way of framing history by reconfiguring its relationship to memory. Genial and outgoing, he sensed that he was riding the crest of a new wave in historiography. Ricoeur, by contrast, was a rear-guard philosopher. By nature, he was modest and retiring. He followed new trends in historical writing with interest, but commented on them only after they were solidly established. He took up the "new" work of the Annales school, for example, as it settled into academic institutionalization.[26] His following project on the theory of narrative appeared long after academic fanfare over the rhetorical turn in scholarship had quieted down.[27] Accordingly, his study of the relationship between memory and history, his last major work, was offered as a mature reflection on decades of scholarship.[28] He probed the meaning of this line of scholarly inquiry

---

[26] Paul Ricoeur, *The Contribution of French Historiography to the Theory of History* (Oxford, UK: Clarendon Press, 1980), 7–12.

[27] Paul Ricoeur, *Temps et récit* (Paris: Seuil, 1983–1985), 3 vols.

[28] Paul Ricoeur, *La Mémoire, l'histoire, l'oubli* (Paris: Seuil, 2000).

as tested against the teachings of the great philosophers of the Western tradition.[29]

Ricoeur challenged Nora for his abandonment of the grand narrative of *mémoire/histoire*. He addressed Nora's thesis about the mnemonic turn from the vantage point of the most neglected realm of his project—that of the relationship between memory and history in Holocaust studies. Whereas Nora had devoted his energy to issues about the French identity, his German counterparts had been intensely preoccupied with the repression of the Holocaust in postwar memory. By the late 1980s, Holocaust studies had become a major field of scholarly interest in Germany, the USA, and Israel. More indebted to Freud than to Halbwachs in method, Holocaust scholars addressed memory as a process of mourning and noted limits to the historical representation of the suffering of victims of Nazi crimes against humanity. Ricoeur, therefore, called upon Nora to take into consideration this neglected line of inquiry, for it revealed the nature of memory's autonomy vis-à-vis history.[30] Here unrequited memory (*insolite mémoire*) bodied forth its claims, for trauma does not respond to the historians' interrogation in the transparent way that Nora proposed. The living memory of the Holocaust remained opaque to the outside world, its meaning held fast within the psyches of its victims. What such memories might reveal was enshrouded in their existential suffering. Before proceeding to issues of identity, Ricouer argued, history had first to beg pardon of these memories. Mourning takes time to come to terms with the ordeal of persecution.[31] Ricoeur, therefore, underscored the importance of sorting out the properties of memory vis-à-vis those of history as modes of evoking the past. Nora may have argued for the unraveling of the memory/history relationship in the present age. But in history's interrogation of memory in the *Lieux* project, Ricoeur asked, was history not laying claim to domination over memory? Was Nora not minimizing the significance of repressed memory, dormant but alive, by limiting his discussion to residues of memory that no longer animated contemporary conceptions of the past?

To adduce his argument, Ricoeur turned to philosopher Jacques Derrida's writing about "Plato's pharmacy," based on a critical reading

---

[29] Paul Ricoeur, "Mémoire: approaches historiennes, approche philosophique," *Le Débat* no. 122 (November-December 2002), 54–55.

[30] Ricoeur, *La Mémoire, l'histoire, l'oubli*, 529–34.

[31] Ibid, 648–50.

of Plato's Socratic dialogue *Phaedrus*.[32] Therein Socrates had raised the question: is the written word as opposed to its oral expression a remedy or a poison in the pursuit of knowledge? Ricoeur's point is that neither memory nor history can impose its will upon the other. Each has its own vocation as a way of truth, yet in ways that can never be totally reconciled. History affirms the reality of the past in fixing its events accurately. It aspires to tell the truth about what happened. But history is always an interpretive construct. It explains the meaning of the past in a way that is self-limiting. History composes a set narrative out of the many ways in which a story might be told. Memory, by contrast, is the seat of the present imagination. It is born of experience, and it is a "little miracle" in its open-ended capacity to awaken the imagination of the past in the present. Whereas history is deliberative and studied, memory is dynamic and inspirational. Memory, therefore, cannot be trapped in the logic of historical narrative. Memory may take unexpected turns, thanks to sudden promptings that revitalize the meaning of the past for the present.[33] Nora took Ricoeur's gentle prodding seriously. It became the basis for their scholarly exchange in a colloquium convened in November 2002, whose proceedings were subsequently published in the journal *Le Débat*. Ricoeur's attention to the methods of Holocaust scholarship on memory also prompted Nora to take into consideration for the first time his own Jewish heritage. Scion of an assimilated Jewish family, he had thought of that heritage as only a marginal aspect of his identity, even though he as an adolescent during the war years had outwitted agents of the Gestapo when they had come to arrest him. Here he was pressed as well by his biographer François Dosse, who sought to find a larger place for Nora's Jewish heritage within his life story.[34]

The terms of the debate between Ricoeur and Nora were couched in an exchange about the meaning of the work (*travail*) of remembering as opposed to the duty (*devoir*) to remember. These terms were metonyms, signatures of opposing ways of thinking about memory. Ricoeur's "work of memory" denoted the Freudian task of acknowledging guilt for past failings (more specifically France's complicity in the Holocaust during the Vichy era), followed by a process of "working through" repressed

---

[32] Jacques Derrida, *La Dissémination* (Paris: Seuil, 1972), 71–197.

[33] Ricoeur, *La Mémoire, l'histoire, l'oubli*, 175–180, 525, 644–46.

[34] Ibid, 51–56; Dosse, *Pierre Nora*, 10, 256–57; Pierre Nora, *Esquisse d'égo histoire* (Paris: Desclée de Brouwer, 2013), 46–49.

memories that continue to haunt present-day consciousness. Nora's approach, by contrast, battened on the Halbwachian notion of the role social power had played in the selection of the main elements of the modern French national memory. His study illuminates history's judgment on the relative importance of artifacts, mementos, and souvenirs of a past that had lost its relevance for defining identity in the present age.

Nora responded to Ricoeur with tact. He took issue with Ricoeur for a critique that aspired to take the memory/history relationship into a realm beyond time when in fact what was most important for understanding his argument was historical time itself. Nora underscored the importance of the context in which the memory phenomenon had emerged—as a transition between "regimes" of historical time. He pointed out that he had embarked on this project in an open-ended way, never quite sure where his research would lead. Along the way he found that his interest in the question of memory vis-à-vis history was an intuitive response to a shift in the contemporary understanding of historical time. The notion of places of memory as a concept for historical interpretation, he contended, is no more strange than the well-established ones to which historians ordinarily have recourse, such as fact, cause, structure, or mentality. At issue, he continued, was a breakdown in a tradition of historical thinking about a relationship between past and present that dated from the time of the French Revolution. Historical realities in the late twentieth century were visibly different from those earlier in the century. Accordingly, the *devoir* of memory as an imperative for the present age had been incited paradoxically by a perceived need to preserve a far greater range of mementos of the past than ever before, all because of uncertainty about what posterity might want to recall.[35]

Nora therefore posited crossing historiographical ideas that were reshaping thinking about the relationship between memory and history.[36] The first was the notion that time is accelerating. This perception was born of the fast-moving pace of innovation in all spheres of life in recent decades, especially in economic and cultural endeavor. As a historiographical interest, politics had been marginalized. Consumerism, gender reconfiguration, and especially technologies of media in a digital age drove this perception. Combined, they reinforced the sense of the irrelevance of the

---

[35] Pierre Nora, "Pour une histoire au second degré; réponse à Paul Ricoeur," *Le Débat* no. 122 (November-December 2002): 24–31.

[36] Nora, "Reasons for the Current Upsurge in Memory," 1–8.

politics of the past for understanding the present age, now cut adrift from a timeline long identified with the modern era. The teleological notion of progress inherent in such thinking had lost its appeal. Not only had expectations for the future become uncertain. So too had the meaning of the heritage of the past. The obsession with memory was symptomatic of the anxiety that sprang from such uncertainty.

The second historiographical idea, Nora proposed, concerned the "decolonization" of history. This concept had less to do with historical time, more with social discontent. With the dissolution of the grand narrative of history as the story of the rise of the nation-state, a surge of memories held by disenfranchised groups welled forth to demand history's recognition. Their cause, too, became a *devoir* of memory in light of new thinking about the politics of identity. The collective memories of these groups had the distinction of being "living memories," as opposed to what were now regarded as faint images of the one that had once promoted a unified conception of French history. Identity politics aroused a rivalry among conceptions of history advanced by particular groups. This combined challenge—the call to memorialize the heritage of the past in all its myriad forms, together with the demand to acknowledge the living memory of minority groups—threatened to take possession of what had once been regarded as the authoritative role of professional historians: their singular ability to interpret the meaning of the past for the present generation. Whereas it was once assumed that history would reign over memory by subjecting it to rigorous critical analysis, memory now made its claims upon history.

Laurent Gervereau: "Pourquoi Canoniser Pierre Nora"

Another type of critique was more practical. So celebrated has been Nora's accomplishment that some critics have asked whether its importance was being exaggerated. Laurent Gervereau, a leading French public intellectual, has cautioned against an overly ready reliance on Nora's thesis. He pointed out the perils of conflating the *phénomène mémoire* with the *moment de Nora*. Acknowledging the value of Nora's contribution to rethinking French history, he wanted to show that no thesis is invulnerable to criticism.[37]

---

[37] Laurent Gervereau, "Pourquoi canoniser Pierre Nora?" *Le Monde.fr/idées/articles/*1 November 2011, consulted 14 July 2013.

Gervereau noted two dangers in Nora's approach to history, one pedagogical the other scholarly. In terms of pedagogy, he claimed that Nora's project has little to offer students coming of age today. It is too much, he argued, to ask beginning students in French history to find an orientation amidst the kaleidoscopic world of places and images that Nora offers as a framework. Chronology matters, Gervereau reaffirmed. More importantly, today's students live in a vastly different world from that contained in Nora's conception of the French national memory, which for all of its innovation is retrospective in outlook. Nora has little to say about the larger world in which the French now live. Our references today, Gervereau pointed out, are not those of Lavisse's nation-state, now disassembled, inventoried, and redeployed for reflection, but rather local and global places of memory that must be interpreted with the future in mind. Nora's history envisions not new beginnings but the recomposition of old ways of understanding the French past. His project offers no pathway into this emerging global culture. Our need today, Gervereau asserted, is to interpret the meaning of the past less with an eye to our present "era of commemoration," more with one looking to the future.

As for the scholarly plane, Gervereau contended that Nora's map of history as a mnemonic landscape renders history vulnerable to identity politics. Far from making history the master of memory in its investigative interrogations, history has become a prey to the biased interpretations that are generated at the topical reference points on his maps of memory. With his emphasis on the rhetoric as opposed to the evidence of history, Nora discounts the honest labors of historians in their primary research. The truth of history lies in hard evidence of empirical facts gathered, more than in representations of the past as it was once imagined. The old method of picking up on traces of evidence is still our best route to knowledge of the past reliable enough to serve as a basis for reflection about our present choices. Nora's forte was his work as a publicist, building a network of useful professional associations and writing critical essays of a theoretical nature. To the limits of history, Gervereau suggested, must be added the limits of memory as an avenue for investigating the meaning of the past, for it recycles old ways of assessment when our need is for new beginnings. Gervereau seems to be saying that historians should rescue their work from this idealist realm of imagined communities in favor of the empirical one of tangible realities. Nora's project, he concluded, is at best transitional. It marks the end of an era in its deconstruction of a once valued model of history. But it provides no guidance for how we may replace it.

NORA'S REPLY TO HIS CRITICS

Nora has taken much of such criticism to heart. He has his own nostalgia for what has been lost in the mnemonic turn in French historiography. His recent writings betray some unease about the Pandora's box that he has opened, for memory, with its present mindedness, has come to trouble history in a new way. He acknowledged with irony that a new historiography that had discarded ideology in favor of rhetoric has opened the way for the return of ideology in a new guise, one of a nature that has politicized the meaning of the past to serve present-minded purposes of particular groups whose interests were once subordinate to those of the society at large. Their varied claims upon the past have had a subversive effect on any quest for interpretative coherence.

These concerns led Nora to think again about why the Positivist tradition of nineteenth-century historiography had held its grip on the interpretation of the French past for so long. Even the Annales movement, which gravitated toward a cosmopolitan viewpoint, never abandoned its French point of departure. In a talk at Blois in 2011, then published in *Eurozine*, Nora reflected on what had been lost with the demise of French history conceived in the old way as *mémoire/histoire*.[38] In that guise, historical scholarship had contributed to the stability of a long tradition of national identity. Before the rhetorical turn with its scrutiny of metanarratives, historians had been able to concentrate on problems of evidence. Historians prided themselves on the long hours they labored in the archives. Positivist in their method, they concentrated on descriptive certainties and contented themselves with telling the stories that unfolded from them. They stood apart from the political fray. They thought of their role as essential in the education of the nation to civic purpose. Research specialists, some, nonetheless, wrote general works for the instruction of students in primary and secondary schools.

Historians in that tradition, Nora noted, played a role of leadership within an emerging profession. Some spoke of themselves as artisans supervising the work of journeymen laborers in the archives. Marc Bloch, beloved among the Annalistes for his scholarly integrity and personal courage, famously cast himself in just such a role.[39] Historiography was

---

[38] Pierre Nora, "Recent History and the New Dangers of Politicization," www.eurozine. com (24 November 2011), consulted 16 August 2013.

[39] Marc Bloch, *Apologie pour l'histoire ou métier d'historien* (1949; Paris: Armand Colin, 1993), 69–79.

scrupulously devoted to problems of evidence, patient sifting through ordinary documents, while never overly hopeful of uncovering the extraordinary find. Positivist historians in the tradition of Lavisse were self-assured in their conception of their role. They were oriented toward the past and wary of moving too close to the present for fear that living memory might cloud their judgment. They took seriously their pedagogical role. They respected the grand narrative of the history of France, so that youth might share a common appreciation of their heritage. The French Revolution was the foundational event of modern history. Adversaries about its meaning were paradoxically bound together in a unified framework of historical understanding.

Not anymore, Nora allowed. Memory has been appropriated by a consumerist mentality about its uses, and as such has come to be enlisted in the service of political causes that distort the historical record. He expressed his particular misgivings about interpretation identified with "neo-colonial" historiography, for he worried that the exaggerated place it ascribed to imperialism in the French heritage would diminish the significance of all that French civilization has been when viewed in its ensemble. His first book had dealt with the Algerian crisis, and one might argue that as a lycée professor in Oran during the late 1950s, he was a late exemplar of France's *mission civilisatrice* in the last redoubt of its far-flung colonial empire.

In this closing assessment, Nora reminds us that history is not what it used to be, and at a time in which the French need the historians' impartial judgment more than ever before. If Nora's project is a requiem for a past approach to history more than a plan for how it should be addressed in the future, it is, nonetheless, an honest effort to separate what is living from what is dead in the legacy of the French past. Such has been the *devoir* that has defined his contribution to our understanding of the relationship between memory and history in our times.

CHAPTER 3

The Politics of Commemoration

COMMEMORATIVE REMEMBRANCE IN THE CULTURAL POLITICS OF THE NATION-STATE

Remembrance of modern war, for its mythologies as much as for its realities, is a matrix to which much of the study of the politics of commemoration has been devoted. Research on commemorations, their modes, and their politics played a major role in defining the field, beginning in the late 1970s.[1] Though such inquiries have expanded and diversified over the past generation, the interest in war remains at the heart of this enterprise, an exploration of the deep disillusionments that dashed the hopes and dampened the enthusiasms of nations at war in the modern era.[2] The world wars of the first half of the twentieth century especially were crucibles of memory for the emotions they generated concerning the loss of millions of lives, the destruction of cities, and most enduring, the psychological scars carried by survivors, soldiers, and their families alike. As historian Jay Winter has observed, memories of the world wars of the twentieth

---

[1] An early example of this genre is Maurice Agulhon, *Marianne into Battle: Republican Imagery and Symbolism in France, 1789–1880* (Cambridge: Cambridge University Press, 1981).

[2] John R. Gillis, ed., *Commemorations; The Politics of National Identity* (Princeton: Princeton University Press, 1994), 150–211.

century cast long shadows. The effects of war remain deeply ingrained in the imaginations of those touched by the experience.[3]

The mnemonic modes of national remembrance of modern wars lent themselves well to historical analysis, as living memories of them began to fade. From the practical perspective of professional scholarship, such studies in the politics of commemoration have appealed to historians for the certainties they promise to report about commemorative practices themselves. While collective memory may be elusive and commemorative rhetoric tendentious, historians recognized that they could systematically inventory and describe the practices themselves—the monuments, museums, eulogies, rituals of commemoration, and iconic pictorial representations of martyrs and heroes. While memories evolve, these artifacts remain anchored in fixed times and places. The interpretative interest lies in explaining how these places of memory were invested with changing meanings over time, particularly if they became objects of contested identity in light of changing constellations of political power.[4]

It would be impossible in a chapter of this length to inventory, let alone analyze, the specialized contributions to scholarship on the politics of commemoration. Such studies are now legion, most of them devoted to the role of official remembrance in the making of public identities. They draw attention to the commemoration of dramatic events or celebrated personalities. Early topics especially favored by historians include memorials of World War I, remembrance of the American Civil War, and iconic personalities, such as Abraham Lincoln.[5] But the list now extends far beyond, reaching into the private memories of ordinary veterans and their families who held on to personal souvenirs.[6]

---

[3] Jay Winter, *Remembering War; The Great War Between Memory and History in the Twentieth Century* (New Haven: Yale University Press), 3, 178.

[4] See John Bodner, *Remaking America; Public Memory, Commemoration, and Patriotism in the Twentieth Century* (Princeton: Princeton University Press, 1992), 13–38.

[5] Examples include Daniel J. Sherman, *The Construction of Memory in Interwar France* (Chicago: University of Chicago Press, 1999); Barry Schwartz, *Abraham Lincoln and the Forge of National Memory* (Chicago: University of Chicago Press, 2000). Jay Winter, *Sites of Mourning; The Great War in European Cultural History* (Cambridge: Cambridge University Press, 1995), and Allen Douglas, *War, Memory, and the Politics of Humor* (Berkeley: University of California Press, 2002).

[6] Noteworthy are Michael Kammen, *Mystic Chords of Memory; The Transformation of Tradition in American Culture* (New York: Knopf, 1991); W. Fitzhugh Brundage, *The Southern Past: A Clash of Race and Memory* (Cambridge: Harvard University Press, 2005); Peter Homans, ed., *Symbolic Loss; The Ambiguity of Mourning and Memory at Century's End*

Here I wish to review some classic studies that first shaped our understanding of the ways of remembrance as a legacy of modern wars. Their interest in memory is closely allied with the study of nationalism in whose names the world wars of the twentieth century were fought. I have chosen six authors whose interpretations center on the crossroads where nationalism, myth, and memory converge.

Imagining and Inventing Tradition: Benedict Anderson and Eric Hobsbawm as Models

Benedict Anderson. I begin with two accidental historians of memory, Benedict Anderson (b. 1936) and Eric Hobsbawm (1917–2012). Both wrote books about the workings of nationalism as a modern ideology. In the process, both coined sententious phrases that caught the eye of scholars embarking on projects dealing with the study of collective memory: Anderson the idea of the "imagined community"; Hobsbawm the concept of the "invented tradition." Both of their books appeared in 1983 and both would soon become required reading for scholars entering the field of memory studies. Anderson reports that his book was translated into 29 different languages.[7] Terence Ranger, Hobsbawm's co-editor, notes that *The Invention of Tradition* was cited in the bibliography of every application to granting agencies in the social sciences in the USA over the decade following its publication.[8] Beyond their expectations, these authors prepared the way for a shift in scholarly interest from ideologies that anticipated the future toward collective memories that mourned the past. The shift served as a basis for rethinking cultural history in the late twentieth century.

Anderson's study is significant for explaining the preconditions that made possible the idea of the modern nation-state as an imagined community. The key, he explains, lies in the transition from dynastic monarchies to nation-states over the course of the early modern era. The dynastic

---

(Charlottesville: University Press of Virginia, 2000); David W. Blight, *Race and Reunion; The Civil War in American Memory* (Cambridge: Harvard University Press, 2001).

[7] Benedict Anderson, *Imagined Communities*, rev. ed. (1983; London: Verso, 2006), 207.

[8] Terence Ranger, "*The Invention of Tradition* Revisited," in *Legitimacy and the State in Africa*, ed. Terence Ranger and Megan Vaughan (London: Palgrave, 1993), 62–63.

state represented a late expression of government conceived as a politics of families. The king was father of his subjects, and his power over his realm was contained in that notion. Bloodlines were important; kinship mattered in defining the echelons of the political and social hierarchy. The people over whom the king reigned often hailed from unrelated places, a patchwork of heterogeneous communities loyal to his person. This congeries of communities of different sorts shared an allegiance to the king as sovereign, and not much else.

The regicide of France's Bourbon family ruler in the French Revolution signaled the death knell of the politics of families, while simultaneously a new political ideal of the general will of the people, given philosophical expression by Jean-Jacques Rousseau, was affirmed in the civic festivals of a new republic. The king's subjects were reborn as the nation's citizens.[9] The transition marked by this radical upheaval may have appeared dramatic. But it was made possible, Anderson contends, by the slow but sure democratization of print literacy, which provided a widening public with the intellectual tools needed to participate in a newly imagined, far more abstract conception of community. The new nationalism was a mythic idea, localized in holidays and festivals, and inculcated in primary school pedagogy. It permitted citizens to adopt a new civic identity and to participate in projects advanced in its name. The nation so conceived came to be grounded in its commemorations.[10]

Anderson's interest in this topic came via his analysis of nationalism's relationship to Marxism. For Marxists, Anderson explains, nationalism was a problematic anomaly on the way to the proletarian revolution that would usher in a classless society. Here, Anderson contends, Marxists failed to grasp the power of nationalism, especially from the vantage point of the realities of the twentieth century. In the midst of the uncertainties of a rapidly modernizing civilization, nationalism surged.[11] If it could not fulfill the promise of social perfection, it could at least provide collective security as a consolation. Accordingly, Anderson argues, the appeal of nationalism lay in its claim to profound origins.[12] Hence the importance he attributes to heritage as the binding sinew of nationalist sentiment. Modern nationalism battened on a

---

[9] See also Lynn Hunt, *The Family Romance of the French Revolution* (Berkeley: University of California Press, 1992), 1–16.

[10] Anderson, *Imagined Communities*, 37–46, 61.

[11] Ibid., 3–4.

[12] Ibid., 67–82, 187.

new conception of historical time, or one might say of a consciousness that transcends it. The nation was thought to embody a kind of consciousness shared by the living and the dead. In this way, heritage implied continuity between past and present in a common social ideal, conceived not as a linear sequencing of time but rather as a belief in the simultaneity of past and present. The nationalist ideal conjured up visions of shared landscapes and shared heritage. These imagined settings in space and time tended to employ stereotypical images. The heroes of national liberation likewise acquired iconic form as they assumed mythic stature in popular recollection. Modern nationalism was powerful, Anderson concludes, by virtue of its appeal to collective memory.[13]

Eric Hobsbawm. Similarly, Eric Hobsbawm's notion of an invented tradition was taken up by scholars in ways that quickly outran his intended use of the concept. Hobsbawm had wanted to show how nation-states of the late nineteenth century, in the pride of their expanding governmental role at home and their imperialist ventures abroad, publicized the deep roots of their national identity in immemorial tradition, when in fact these roots were shallow where they existed at all. He took pains to distinguish newly invented traditions from older authentic ones hewn out of custom through centuries of practical improvisation.[14] The invented tradition, he argues, was not based on precedent but rather on a selective and idealized representation of the past conjured up to serve the present-minded purposes of the nation-state.[15] These invented traditions provided much needed cultural cohesion for a civilization in rapid transformation. The cultural props of the old ways were fading fast. Church and monarchy no longer inspired faithful allegiance in the way they once had. Political power was increasingly centralized, government administrations grew in their outreach, and mass electorates came into being to demand a participatory role in the selection of their leaders. Newly invented traditions fostering patriotism played a crucial role in building allegiance to the new national centers of power. The conscious appeal to tradition strengthened emotional bonds between elected officials and their constituencies. Expanding public

[13] Ibid., 192–99.
[14] Eric Hobsbawm, "Introduction: Inventing Traditions," in *The Invention of Tradition*, ed. Eric Hobsbawm and Terence Ranger (Cambridge: Cambridge University Press, 1983), 2–3.
[15] Ibid., 4–14.

systems of primary education were enlisted in the project of inculcating civic pride and the responsibilities of citizenship in the young. National holidays were instituted or refurbished to punctuate the calendar of what was in effect a new secular religion of nationalism. National flags became sacred emblems. Imposing monuments to epic historical events became salient commemorative reminders of the nation-state reconceived as the community of primary allegiance. The format and content of these practices varied from country to country, but the instruments for fabricating and sustaining the new cultures of nationalism were everywhere much the same.[16]

Hobsbawm's thesis was meant to be provocative. But the scholars' uses of the concept soon ranged beyond the scope of his interpretation. By the late 1980s, the new realities of an age of economic and cultural globalization had displaced the old ones that had given nationalism its considerable appeal a century before. Nationalism as an ideology, moreover, had become suspect in light of the devastating wars of the twentieth century carried out in its name. The concept of the invented tradition exercised a compelling appeal to scholars, I would argue, because traditions invented to buttress the authority of the nation-state no longer spoke to the needs of the present age. Scholarly interest, therefore, shifted from tradition's ideological appeal to the politics underpinning its construction. In these newfound circumstances, many readers were willing to believe that any and all traditions were invented to serve tendentious political ends, lending a cynical cast to the idea of tradition itself.[17] The autopsy of tradition, therefore, became the working model for scholars taken with Hobsbawm's stimulating study.

VARIATIONS ON THE TOPIC OF COMMEMORATION

What follows are variations on approaches to national commemoration. I offer précis of the arguments of four authors. Each takes a different tack: George Mosse, the central role of commemoration in the making of the German national identity; Yael Zerubavel, the amalgamation of ancient and modern sites of memory in the construction of Israeli national

---

[16] Hobsbawm, "Mass-Producing Traditions: Europe, 1870–1914," in ibid, 263–307.

[17] See the critique by Mark Salber Phillips, "What is Tradition When it is Not Invented? A Historiographical Introduction," in *Questions of Tradition*, ed. by Mark Salber Phillips and Gordon Schochet (Toronto: University of Toronto Press, 2004), 4–8.

identity; Jean-Marc Largeaud, the interplay of memory and history in the French commemoration of the battle of Waterloo considered over the long run; and Jay Winter, historical remembrance as a hybrid of memory and history in conveying the meaning of World War I.

GEORGE MOSSE ON THE MYTHOLOGY IN GERMAN NATIONALISM

George Mosse (1918–1999) in his long and distinguished career as cultural historian studied the trajectory of nationalism from ideology to commemoration over the course of modern German history. He was among the first to interpret the role of commemorative practices in German nation-building. Emigrating from Germany to the USA as a young man, he was educated at Swarthmore and Harvard.[18] He had an ongoing interest from the late 1950s in the ideological roots of National Socialism, which he traced to a populist conception of nationalism grounded in an imagined German rural landscape and a mythologized past.[19] German nationalism assumed an idealist cast, he argues, because of the historic heterogeneity of the myriad German-speaking states and principalities of central Europe, and the long and halting task of German unification under Prussian auspices over the course of the nineteenth century. Put more succinctly, Germany was an idea long before it became a nation-state. Its mythological conceptualization by writers and philosophers harked back to a deep cultural heritage identified with the attitudes and beliefs of the German people (*volk*) in a highly idealized representation of their past. This myth of Germany as a people who from antiquity shared a common consciousness took on new political meaning in the modernizing campaigns of Prussian statesman during the Wars of Liberation (1813–1814). German nationalism came to be identified closely with these campaigns. The victory over Napoleonic forces would anchor a legendary history, harking back to the victory of Hermann and

---

[18] See Mosse's autobiography, *Confronting History; A Memoir* (Madison: University of Wisconsin Press, 1999).

[19] George L. Mosse, *The Nationalization of the Masses; Political Symbolism and Mass Movements in Germany from the Napoleonic Wars Through the Third Reich* (New York: Howard Fertig, 1975), 1–20.

his horde of Germanic warriors in their campaigns against the Roman legions during the first century CE. Because German claims to a national identity were so ethereal, bound more to the cultural mainstays of language, ethnicity, and mores than to political institutions, nationalist statesmen portrayed soldiers in these modern wars as heroes reenacting the struggles of their ancestors in defense of their native land.[20]

Like Anderson, Mosse contends that the rise of nationalism at the turn of the nineteenth century was a response to the decay of senescent social and political institutions dating from the Middle Ages. Nationalism filled a need for a broadly conceived idea of community that refashioned the old notion about a German collective unconscious in a new ideological guise. In principle, the new nationalism championed a kind of egalitarianism, not of means but of mindset, or as sometimes professed in loftier terms, a collective soul. In this respect, German nationalism also drew upon Christian religious sentiment, notably notions about an inner voice of the sort associated with Pietism. This suggests why the new nationalism may be interpreted as a civic religion.[21]

At the same time, Mosse explains, nationalism was an ideology of considerable ambiguity. It was at once radical and conservative—radical for the activism it sought to generate, conservative in its emphasis on cultural rootedness in homeland and in heritage. The new nationalism had its high priests: professors and writers such as Johann von Fichte and Ernst Arndt, and activists such as Friedrich Jahn, famous for his role in the formation of athletic and fraternal youth societies. They idealized the vitality of youth, for youth movements were essential to the image of the new nationalism. Gymnastic societies, male choirs, and sharpshooters were mainstays of nationalist ventures throughout the nineteenth century. Such fraternal societies appealed to the idealism of the young themselves. They fostered camaraderie, shared activism in the service of a cause, emotional bonding, an outlet for youthful energies, particularly in sporting activities. They also offered an escape from the routines of daily life in the promise of adventure in defense of the fatherland.[22]

As Mosse remarks, so abstract a notion of nationalism sustained its appeal through the aesthetic design of its commemorative practices. These were fashioned to reinforce remembrance through images that

[20] Ibid., 21–46; Mosse, *The Crisis of German Ideology; Intellectual Origins of the Third Reich* (New York: Grosset & Dunlap, 1964).

[21] Mosse, *Nationalization of the Masses*, 73–85.

[22] Ibid., 127–33.

glorified the nation in both space and time. The sacred space of German remembrance was the landscape, the fields and forests in which its people had drawn emotional sustenance since time immemorial. In a world of urbanization and industrialization, nationalists found solace in nostalgia for a vanishing rural way of life. Writers and artists from an emerging middle class idealized the common man who tilled the soil of German farmland in the manner of their forefathers. The new nationalism had its sacred time as well. Nationalists proclaimed the primordial origins of their cause. They showcased German heroism, notably in war. Ancient military battles were juxtaposed to modern ones. A newly constructed monument to the victory of the Germanic chieftain Hermann over the Roman legions (9 CE) was venerated as a place of memory as important as that commemorating the battle of Leipzig (1813) that capped the Wars of Liberation in the early nineteenth century. The creation of commemorative statuary remained a mania anchoring the cult of remembrance throughout the nineteenth century.[23]

Mosse's perspective on German nationalism evolved over the course of his scholarly career. His last, and perhaps best written work, concerned the formation of the cult of the fallen soldier during World War I.[24] Here the rhetoric of German patriotism once voiced by enthusiasts for war took on a mournful tone in coming to terms with military defeat and the fall of the German Empire. The exaltation of heroic youth gave way to subdued meditation on soldiers who had sacrificed their lives for the fatherland. The image of Germany as an untamed forest was domesticated in the pastoral settings of the military cemeteries constructed to house the war dead. Spare, uniform, elegant in their simplicity, these places of memory rerooted the nationalist ideal in this hallowed ground. Mosse labeled such elegy the Myth of the War Experience.[25]

Despite the postwar zeal for commemorations, the cult of the fallen soldier could not sustain the emotions roused by war indefinitely. In time, memories of the sacrifices of combat veterans were transmogrified in two ways. First, remembrance of the war came to be trivialized in the sentimentality of war souvenirs. Such kitsch included postcards, toy soldiers, parlor games, and battlefield tourism in a comfort that contrasted dramatically

---

[23] Ibid., 47–72.
[24] George L. Mosse, *Fallen Soldiers: Reshaping the Memory of the World Wars* (Oxford: Oxford University Press, 1990).
[25] Ibid., 6–8.

with the hardships of those who had gone to war.[26] Second, and more disturbing, was the corruption of the myth of nationalism, turned to extremist political ends. Nationalism during the 1920s and 1930s regressed into crude aggressiveness with the appearance of a new kind of "volunteer." He was no longer the idealistic youth who had signed up for service at the outbreak of hostilities, but rather a war veteran hardened by its brutalizing and senseless campaigns, now frustrated by defeat, numbed and coarsened by its violence. At loose ends, some veterans formed Free Corps, the prototype of the extremist paramilitary organizations that set out to intimidate the leaders of the Weimar Republic. Apologists such as Ernst Jünger portrayed them as exemplars of a new race of men, warriors emboldened by the realities of war to revive a defeated nation through vigilante action.[27]

It was in this political climate, Mosse argues, that National Socialism found fertile ground. Hitler took advantage of the resentment of a defeated nation, and turned it toward his racist political ends. Hatred of an imagined enemy—the Gypsy, the homosexual, and especially the Jew—played into popular emotions in visceral ways. Vitiated by the Nazi crimes of genocide, the Myth of the War Experience after World War II was enshrouded in shame and so could not resuscitate the cult of the fallen soldier as it had been venerated in the immediate aftermath of World War I. The memory cycle of the myth sustaining the new nationalism in Germany had run its course.[28]

YAEL ZERUBAVEL: ZIONISM AND THE POLITICS OF COMMEMORATION

Intriguing as a comparison with Mosse is the book by the Israeli–American sociologist Yael Zerubavel, *Recovered Roots* (1995), a study of the revision of Jewish collective memory by leaders of the Zionist movement in the early twentieth century.[29] A sociologist by training, Zerubavel is professor of Jewish Studies and History at Rutgers University. Her method is exemplary as a model in this genre for her explanation of the way national

---

[26] Ibid., 126–56.

[27] Ibid, 159–81.

[28] Ibid, 201–20.

[29] Yael Zerubavel, *Recovered Roots; Collective Memory and the Making of Israeli National Tradition* (Chicago: University of Chicago Press, 1995).

memory is created and refashioned over time. She shows how leaders of the newly created nation-state of Israel constructed an official heritage by juxtaposing widely removed and unrelated episodes in Jewish history, two ancient (Masada, Bar Kokhba) and one modern (Tel Hai). She traces the evolution of Israeli national memory from sacred to profane conceptions in a politically charged cycle—from veneration of these episodes to their comic deflation once the foundations of this fledgling nation-state were secure. Her study suggests that the perennially popular notion that history moves in cycles actually concerns the cyclical dynamics of the cultural recourse to memory.

Zionists, Zerubavel explains, proposed to return to the land of their Jewish ancestors, from which they had been expelled nearly 2000 years before. There they planned to rebuild that ancient nation anew. The comparison of German and Zionist nationalism is not without irony. It was in response to anti-Semitism in central Europe during the late nineteenth century that Jewish leaders took initiatives to form a nation of their own. The Zionist movement out of which the Republic of Israel would emerge after World War II is especially interesting because of the nationalist zeal of its activists and the speed with which it succeeded in reestablishing a Jewish presence in Palestine during the first half of the twentieth century. As they staked their claim to what had long since become a strange and alien land, Zionist poets, writers, and political activists of the early twentieth century turned to the task of constructing an imagined community fashioned in remembrance of an ancient heritage.[30]

Zerubavel takes seriously Hobsbawm's idea of the invented tradition. Zionism was a vision of a new nation that longed for oldness. Her study traces the way a newly founded state builds a cultural identity. She reviews its fortunes from the settlements of Zionist pioneers of the prestate period through the wars in which Israel established and then defended its identity as a nation-state. In many ways, Zionism was a nationalist movement not unlike its European counterparts, though its beginnings date only from the late nineteenth century in the campaigns of the Austro-Hungarian journalist and activist Theodor Herzl to reestablish a Jewish state in the place of its ancestral heritage. The Zionist movement inspired much enthusiasm among Jewish youth in Europe, and migration to Palestine proceeded apace during the early decades of the twentieth century. But Zionists were establishing a homeland in territory to which they had only the most tenuous

[30] Ibid., 13–15.

modern claim. More than European nationalists, therefore, they depended heavily on the construction of a historical tradition to justify their cause. This was not an easy task, for it required gathering together scattered memories of heroic actions in a distant past and weaving them into a new narrative of Jewish history. In the myth of nationhood so devised, modernity and antiquity were perceived to be allied as corresponding phases within a broadly conceived historical continuum. A nationalist movement with virtually no modern roots revitalized events out of the depths of its Jewish heritage. These recovered roots became the historic places of memory of modern Israel's identity.[31]

Zionists, Zerubavel argues, embarked upon their cause of nation-making with uncompromising conviction. They repudiated the attitudes that had shaped the religious culture of the Jewish diaspora. They looked down upon the exilic Jews of Europe who for 2000 years had yielded in the face of discrimination against them and who were unwilling or unable to resist the persecution to which they were subjected in twentieth-century Europe. Indeed, Zionists defined their identity against what they perceived to be the passivity of Jews in exile. Even after World War II, Zionists were slow to find compassion for victims of the Holocaust in Hitler's Germany. Only decades after World War II did Israeli statesmen seek to integrate its memory into their own conception of national identity.[32]

In formulating a myth of national origins, Zerubavel explains, Zionists of the prestate era radically revised the sacred history of Jews in exile. They repudiated what had been the theological cast of Jewish history conceived as a religious heritage. Exilic Jews had maintained their culture in widely scattered communities through the binding ties of commemorative religious practices that served as the foundation of their collective identity. The defeat and dispersion of their forbearers in antiquity was interpreted as a tragedy, and the wisdom of their prophets and teachers a consolation. Zionism offered a secular alternative, revising Jewish history so as to represent the Exile as an interim period between the nation of Israel in antiquity and its modern Zionist revival. As a reinterpretation of Jewish history, the Zionist narrative was highly selective, and it replaced one tradition of collective memory with another to advance its cause. The notion that modern Zionist pioneers in Palestine in the early twentieth century were

---

[31] Ibid., 15–36.
[32] Ibid., 75–76.

recapturing the energy of their ancestors was essential to the myth of the new Jewish state in the making. Zionists sought to reaffirm their symbolic connections with the courageous deeds of that ancient nation, recalling their fight to the end as they faced obliteration by the Roman legions. They taught that the present age was witnessing the rebirth of that heroic confidence. The passivity of Jews in the long exile would be displaced by the active engagement of their descendants in the new tasks of rebuilding that historic nation. Reframing the collective memory of that heritage, therefore, was vital to the meaning of the Zionist cause. They celebrated their leaders now as avatars of leaders back then.[33]

In an explanation not unlike that of Mosse for German nationalism, Zeruabavel shows how Zionist intellectuals and statesmen juxtaposed remembrance of ancient and modern military actions as key elements in their construction of a new national memory: the battle of Tel Hai in 1920, the revolt of Bar Kokhba in 132 CE, and the last stand at Masada in 73 CE. In the pioneer prestate days of the early twentieth century, Tel Hai was a much celebrated historical event for the courage and spirit of self-sacrifice early settlers displayed in their skirmishes with neighboring Arabs. For this event, Zionist commemoration focused on the death of Josef Trumpeldor, a charismatic veteran from the campaigns of the Russo-Japanese War, who in Palestine became commander of the Mule Brigade under British supervision during World War I. Dying in a shootout while defending his settlement in Upper Galilee, Trumpeldor was reputed to have uttered the edifying last words: "Never mind, it is good to die for the country."[34]

In commemorating the life of Trumpeldor, his memorialists could point to living witnesses to his dying declaration. Memory of the events that transpired at Bar Kokhba and Masada, by contrast, was beclouded by suspect evidence retrieved out of a nebulous past. Neither had figured positively in Exilic tradition, for both were remembered as episodes of failure in defeat. They would, nonetheless, find a restored place in Zionist collective memory because they exemplified the spirit of active resistance against all odds that Zionist leaders wanted to instill within their youth as a strategy for deepening their commitment to the present cause: the heroic revolt under Shimon Bar Kokhba in the face of inevitable defeat

---

[33] Ibid., 22–36.
[34] Ibid., 43–47.

by vastly superior Roman legions; mass suicide at Masada as a courageous alternative to abject surrender to the Romans. Zionists telescoped these events into a mentality shared across the reaches of time.

The stance of intransigent defiance that characterized all three episodes, Zerubavel explains, would become the lore around which the Israeli nation would fashion its culture of remembrance through highly effective commemorative practices. The sacrifices they recalled were integrated into the rituals of a holiday cycle of annual observance. The stories about the heroism they had exhibited became exemplary models for Israeli school children. The historic sites of Bar Kokhba and Masada became places of pilgrimage. The heights of Masada especially, by virtue of their remoteness, served as sacred ground for visitation, first for treks by intrepid youth, eventually for tourism by the public at large. The glue that held these commemorative practices together was the revival of ancient Hebrew as the language of instruction in public schools.[35]

Zerubavel makes a persuasive case for the construction of a collective memory to which nearly all citizens could subscribe during the prestate period of nation-building. For the most part uncritically accepted by settlers, its myths were essential to promoting a sense of shared identity. The interest of Zerubavel's account, however, also lies in her explanation of the way these tales of death-defying heroism were in time challenged and subverted, ironically because the task of nation-building had been so successfully accomplished. As a nation-state from 1948, the Republic of Israel would continue to see itself as a nation besieged by hostile neighbors, and the myths of origins would never be officially abandoned. But the passage from prestate Zionism into Israeli statehood soon revealed the limits of an ideology whose enthusiasms relied so heavily on the affirmation of martial zeal whatever the cost in soldiers and resources.[36]

Zerubavel goes on to show how the unity inspired by reverence for a legendary past dissipated in the decades following nationhood, roughed up by ongoing tensions with hostile neighbors. The bane of nationalism, she points out, is its need for constant reinvigoration. The Zionist myth of origins was periodically resuscitated, as Israel went to war with its Arab neighbors in 1967 and again in 1973. Victories notwithstanding, the wars

---

[35] Ibid., 28–31.
[36] Ibid., 229–37.

exposed Israel's vulnerability, and incited parliamentary debate about the best policies to insure the well-being of the nation. Uncompromising defiance in the manner prescribed in the episodes of historic remembrance could no longer command blind faith. Statesmen debated whether it was not wiser to seek accommodation with Palestinian Arab neighbors by making concessions to their demands to share this tiny land. Given the realities of survival in the midst of present tensions, skeptics asked, was not temporizing statesmanship a better plan for national security than activism in the name of stubborn national pride? The wars of 1967 (Six-Day War) and 1973 (Yom Kippur War) may have been stunning Israeli victories, but they left a legacy of worry about how vulnerable this fledgling nation remained. A once coherent collective memory unraveled into particular collective memories in controversies over public policy as Israel faced its ever precarious situation.[37]

On the intellectual plane, the myth of the Republic of Israel's profound origins, once naively accepted in Zionist collective memory, was deflated by sobering historical doubts about how little one could know about what actually transpired in those places in those ancient times. The veracity of the legends about them was challenged, as historians got into the act. In the process, sacred memories were subverted in these profane reassessments, as the patriotic narratives about these events were subjected to close examination. Historians pointed out the bias of ancient historians, notably the Roman Dio Cassius and the Jew Josephus, on whose accounts memory of these events was based. Shimon Bar Kokhba, leader of the revolt that bears his name, was exposed as a shadowy figure whose identity could not be confirmed in a reliable way. Was mass suicide at Masada, critics asked, the only solution for Jews facing the Roman legions? Even testimony about the exact words in which Trumpeldor issued his dying declaration was questioned, and his words became the butt of subtle humor. Still, the authority of patriotic remembrance of these legendary origins was challenged only in fits and starts, and only to some degree.[38] The Republic of Israel has prospered as a modern nation-state. But to this day, its identity has remained so inextricably bound to its legendary heritage that it troubles its relations with Arab citizens and neighbors in ways that seem insoluble.

[37] Ibid., 190–97.
[38] Ibid, 197–213.

JEAN-MARC LARGEAUD: HISTORY WITHIN THE CYCLE OF COMMEMORATION

A breakthrough study about the evolution of commemorative practices in remembrance of war is Jean-Marc Largeaud's *Napoléon et Waterloo* (2006). The definitive fall of Napoleon in the battle of Waterloo signaled the end of French hegemony in Europe. But its memory in French popular culture, Largeaud explains, was over time transfigured into a "glorious defeat," an emblem of devotion to duty, loyalty, and patriotism invoked to foster national renewal in times of adversity.[39]

Largeaud canvasses nearly two centuries of reflection on its meaning in the politics of French culture. The vanquishing of Napoleon's army at Waterloo signaled the definitive end of French hegemony in Europe. But its memory continued to captivate the imagination of Frenchmen, and over time was transfigured into an idealized image of a courageous last stand. The battle was short and decisive; the telling and retelling of tales about it lingered on for nearly a century. Largeaud's study, therefore, is of particular interest for tracing the poetical logic of commemorative discourse when considered over long periods of time. Waterloo, experienced by its soldiers as the existential suffering of agonizing combat, would come to be remembered by following generations as a disembodied emblem of devotion to duty, loyalty, and patriotism against all odds. In the process, Waterloo came to be known as an oxymoronic "glorious defeat," salvaging heroic remembrance from the collapse of France's most memorable regime. So its lessons would be enshrined in popular histories, the schoolbooks of children, romantic novels, and the ritual ceremonies of national commemorations. In these many ways, the losses of the battlefield found recompense in the psychological consolation of memorable lessons for posterity.

The notion that Waterloo had paradoxically been a "glorious defeat" nurtured popular hope for national regeneration and renewal. In this it drew upon the Napoleonic legend. Had not Napoleon himself returned from exile to take up the cause of restoring the glory of the French nation once more? Napoleon had not prevailed, but the French nation would recover, stronger and more dedicated to national pride than before. Waterloo provided an edifying memory of the resilience of the human

---

[39] Jean-Marc Largeaud, *Napoléon et Waterloo; la défaite glorieuse de 1815 à nos jours* (Paris: Boutique de l'Histoire, 2006), 17–21.

spirit. The memory of Waterloo was invoked by French statesmen and educators, notably after the Second Napoleonic empire suffered defeat at the hands of Germany in 1871. In civic lessons, Waterloo was contextualized within the longer story of French history, a moment of misfortune like others from which the French had recovered over the centuries, harking back to the French defeat at Agincourt in the late Middle Ages and deeper into a mythical past as sung in the story of the death of Roland in the time of Charlemagne.[40] As testimony of sacrifice, military pride, and male honor, remembrance of Waterloo was recast to serve a more abstract purpose, as patriotism became a civic religion for the Third Republic during the late nineteenth century. In an even more abstract image that circulated beyond French frontiers, Waterloo came to represent a universal mindset, a mentality characteristic not only of France as a nation in defeat, but of any such nation at it sought recompense for its failure in the face of adversity.[41]

Along the way, the memory of Waterloo also played into nineteenth-century French politics, each party invoking the battle as it served its purposes. In such historical remembrance, unreliable memory often introduced distortions into its representation. Largeaud shows how testimony by living witnesses soon shaded into hearsay. Details were recomposed in popular narratives, before being conflated into emblematic images. Popular histories bore the mark of their authors' biases. Largeaud points out that even Henri Houssaye's three-volume history of the battle fell short in his efforts to disentangle memory from history. Highly popular among both historians and the public at large, his history went through 45 editions after its publication in 1898.[42] Houssaye professed to write history in the scientific mode of the professional historiography of the day. His work was praised for its erudition by leading contemporary historians, such as Gabriel Monod and Alphonse Aulard.[43] Yet from Largeaud's perspective, Houssaye's history betrayed the bias of his time in history. If his study was learned and authoritative, it was also framed in a way that exalted the making of France as a nation-state.

Much of the interest of Largeaud's study, therefore, lies in the rhetorical strategy with which he has structured the relationship between

[40] Ibid., 163–64.
[41] Ibid., 234–42.
[42] Ibid., 165–70.
[43] Ibid., 167.

memory and history. He proceeds chronologically through the stages of modification, revision, and redeployment of collective memory of the battle—from testimony by participants, to recomposition of their stories in journalistic narratives and political propaganda, to its reconstruction by professional historians at century's end. But then—in what may be his most original perspective—he shows how the history of the event came to be rivaled by its memory once more, this time in the guise of its imaginative re-creation. Here he explores the representations offered by novelists, poets, playwrights, and painters over the course of the nineteenth century, all of whom sought to imagine the battle as it dramatized some deeper issue about the human condition. For this reason, Houssaye's history was rivaled in popularity by the fictional representations of Stendhal and Victor Hugo who, in reaching for the historical sublime, turned history into myth once more. They, and artists, dramatists, and novelists like them, directed attention to more abstract ideals for which the event itself was little more than a convenient prop. The chain of the story's refashioning, Largeaud concludes, reveals how malleable historical remembrance can be. He therefore situates the historiography of the battle between two phases of collective memory—the first about commemoration of the event, the second about the moral imagination.[44] Interest in the battle, he explains, may have been marginalized. But for the student of collective memory, the battle acquires renewed historical interest for the longevity of its afterlife in the varieties of its modes of remembrance. As factual event, Waterloo has its place in the historical record once and for all time. As memory figure, the event lives on.

JAY WINTER AND THE HISTORICAL REMEMBRANCE OF WORLD WAR I

I close this chapter with a few remarks about the work of historian Jay Winter (b 1945) on the commemoration of World War I. He is among its foremost authorities, given the range and complexity of his analysis and the insight with which he relates his findings to the larger topic of the relationship between memory and history. Like Zerubavel, he is sensitive to the ways in which memory and history draw upon shared resources, even as they pursue separate and distinct approaches to understanding the meaning of the past. Reflecting on work on war, myth, and memory

[44] Ibid., 273–343.

(including his own), he implicitly tests the limits of the heuristic concepts of the "imagined community" and the "invented tradition" out of which so much scholarship on collective memory has been drawn. Here I comment only on his most recent book, *Remembering War* (2006), which places all the work on collective memory in relation to historical understanding in a comprehensive historiographical perspective.[45]

Winter is suspicious of the notion of collective memory for its vagueness about who it includes and how it operates. He argues that discussion of the collective memory of a nation is a dubious proposition, tendentious and even mythological in its formulations. As an imagined community, the nation is a flimsy and evanescent structure for remembrance. There are times and places in which shared sentiments of patriotism and national identity may be evoked, he allows. But in recollecting the experience of war, there are many communities of remembrance, and it is in these that memories of war are most deeply implanted. He therefore goes in search of a middle ground between memory and history. Each has its resources for evoking the past. Memory and history as modes of understanding the past are different in nature. But in many ways they interact.

To explain how, Winter coins the concept of "historical remembrance."[46] The study of historical remembrance takes place in that space in which memory and history encounter one another. Collective memory implies passive reflection; historical remembrance calls upon active engagement in the projects of remembering. Following the critic Walter Benjamin, he proposes that we look upon collective memory as a theater in which the past is reenacted. The task is to understand the many and varied practices through which memory is portrayed on that stage. Such practices may be studied concretely. In the case of World War I, these include letters, diaries, plays, novels, movies, even the proceedings of courts of law.[47] All are media through which the experience of the past is given expression. Memory, he explains, is twice filtered. Experience can only be communicated through its representation, and all such representation is selective. It cues what and how experience is remembered. At this point of memory's reception, however, the notion of a collective memory breaks

---

[45] Jay Winter, *Remembering War; The Great War Between Memory and History in the Twentieth Century* (New Haven, CT: Yale University Press, 2006). See also his detailed earlier study, *Sites of Memory, Sites of Mourning; the Great War in European Cultural History* (Cambridge: Cambridge University Press, 1995).

[46] Winter, *Remembering War*, 9–13, 278, 288.

[47] Ibid., 183–86, 275, 278.

down into collected memories. Some people may share common attitudes and images. But as individuals they will never interpret representation of the past in exactly the same way. Memories are too subjective, too much shaped by the varied perspectives of those called upon to remember to be aggregated into a unified conception. In most instances, collective memory is no more than a useful fiction.[48]

Winter contends that the bonds linking individuals in their evocation of the past are more easily recognized in the activities of memorialists, the agents of commemorative practices. A few of them built imposing monuments of national commemoration. But far greater numbers erected more modest memorial structures in small towns and villages. Local committees saw to commemorations by choreographing ritual ceremonies. In such settings, memories of those dear to the community were held fast for personal reflection. Winter, therefore, would have us understand the degree to which commemorative practices are best appreciated for the specific communities to which they appealed. In looking for evidence, one most often finds it on the local level.[49]

Winter also shifts attention from war's heroes to its victims. Following literary critic Paul Fussell, he points out that the primary trope of literary remembrance of World War I was irony.[50] The outbreak of the war had engendered great expectations among young men for the experience of valor in combat. In the trench warfare that followed, however, such notions were dispelled. The rally around the initial call to duty fell apart in suffering on an unprecedented scale. Nearly 10 million soldiers died in World War I, and some 20 million more were wounded. Most extensive but least visible among these wounds was the psychological damage, as survivors were permanently impaired by the shell shock of battle. Combat veterans lived with unrequited memories they could never completely assimilate. Nor were soldiers the only ones who suffered. Mothers, wives, and families were victims as well. All would carry the scars of war as long as they lived. Over the long run, Winter explains, such memories are more likely to be borne most profoundly within families.[51] World War I is best appreciated today as the setting in which the devastating

---

[48] Ibid., 3–6.

[49] Ibid., 150.

[50] Paul Fussell, *The Great War in Modern Memory* (Oxford: Oxford University Press, 1975), 3–35.

[51] Winter, *Remembering War*, 40–42.

psychological trauma of warfare was first acknowledged. That may be why this war remains so prominent in modern memory. From the perspective of cultural history, World War I is significant for the way we have come to mourn its losses rather than to celebrate its campaigns, a precedent for understanding wars yet to come.

Winter therefore questions whether the scholarly focus on the nation is the best venue for understanding the historical remembrance of World War I. Its battles may have been fought in the name of nation-states, and its first memorialists paid most of their attention to soldiers fallen in battle. But the experience of the war was felt in searing ways by combatants and non-combatants alike. The remembrance of World War I, he argues, was enacted on a wider stage, drawing in the many communities touched by its violence and displacements, each in a different way. For Winter, time itself dissolves the coherence of national remembrance, as one traces its fortunes over the long run. While nations in their ideological faith may proclaim long-term continuities between past and present, they change demographically and politically over time and the meaning of national remembrance evolves with them.[52] Living memory is dynamic. It defies the best commemorative efforts to hold its values in place. Even as commemorative practices survive, their meaning undergoes transformative change. Referring to France, Great Britain, and Germany as examples, he notes that the composition of their populations today is far more diverse than it was a century ago. Vast numbers of people migrated in and out of these combatant nations over the course of the following century. Nations changed policies in the face of new realities. The issues that had provoked the outbreak of World War I vanished. Meanwhile, memories of the war lived on among families with considerable staying power. The families who remembered the war and meditated on its losses, Winter speculates, may be thought of as an imagined community spread around the globe.[53] Today, the memory of World War I continues to be culled in a myriad of reflective ways in a variety of settings. For those who meditate on its meaning, its remembrance provides edifying reminders of the wages of warfare.

Winter proposes that the many modes of remembrance in today's world pose a challenge to the historian. The interest in memory as a topic for scholarship encourages historians to use their skills not only to establish a critical perspective on memory's workings, but also to rethink the way

[52] Ibid., 156–66.
[53] Ibid., 168–69.

they themselves work as scholars. Gone are the days in which professional historians could research and write in splendid isolation, should they hope to reach an audience beyond colleagues in their field. The old days of print culture in which historians jealously guarded their individual autonomy has led to their marginalization. They are read by one another, sometimes by their students, but not often beyond. Media is the mode of popular communication today. Television and film reach enormous audiences. The new media of television and film, Winter counsels, should be embraced for the possibilities they offer to renew public interest in the past. In these new modes of communication, the line between memory and history may sometimes blur. But historians would be wise to become engaged in the production of media presentations of history if they wish to exercise their influence on the public at large. Should they fail to do so, those with other agendas will be sure to take up the task.[54]

HISTORIANS OF THE POLITICS OF COMMEMORATION: SOME COMMON DENOMINATORS

As varied as may be the setting to which these historians of commemorative practices have directed their attention, there are some common denominators in their larger interests. Each one focused on the foundations of nationalism, considered in the guise of civic religion. Each sought to expose the politics underpinning commemorative practices that advanced patriotic allegiance to the nation-state through an appeal to heritage rather than statecraft. Each referred to particular historical events to anchor collective memory of the nation's cause, but selectively and tendentiously. Each showed how the aesthetic gloss of commemorative memorials and rituals screened the existential suffering of soldiers on the battlefield. Each arrived at an interpretation that embraced both history and memory in a hybrid genre of its own, which Jay Winter labeled "historical remembrance."

Finally, as an underlying proposition, all of these scholars develop historical perspectives specific to a kind of commemoration identified with the modern age—reverent, eulogistic, promotional in behalf of the nationalist ideal. Commemorations of those who gave their lives in battle invited meditation on the nobility of death in the name of an honorable cause. The historians themselves, of course, distance themselves from such idealized associations and signal the end of a certain kind of commemoration of

[54] Ibid., 289.

war. They reveal how different are the perspectives of the present age, one informed by the "culture wars" of the 1980s over the politics of collective memory. Not that any of them would contend that commemoration of war is about to disappear. The myths attending nineteenth-century nationalism may have lost their force. But the need to pause over the death-dealing wages of warfare is profound in the human condition and in any age inspires meditation on the relationship between the passions of warfare and the reality of human finitude. Commemoration of war today, when it is sincere, is more openly defined if not more lightly taken. The American memorial to the war in Southeast Asia erected in Washington DC is exemplary. It is open to the personal and private interpretation of the individuals who visit and meditate upon it. Protest against that war, particularly among the young, signaled growing skepticism about ready participation in military service out of naive patriotic commitment.[55] The American government conceded this point in its abandonment of the universal conscription of "citizen-soldiers" in favor of building a professional army. Moreover, the ideal of nationalism, long the touchstone of social identity, was being challenged by that of competing allegiances. Political scientist Benjamin Barber has explained the waning appeal of the nationalist ideal in light of a growing awareness of the "end of autarchy," the autonomous nation-state in an age of globalization.[56] Nationalism, as characterized by Benedict Anderson, has moved toward a conception of the nation as one imagined community among many, compressed between local and global places of memory. Insofar as the monuments to the wars among nations of the modern era still hold an appeal, it is more often for their aesthetic effect than for their edifying lessons.

The interpretations offered by these scholars concerning the role of myths among nations at war enable us to understand how collective memory is at once powerful and fragile. It is powerful in the imagination it can quicken and the convictions it can inspire. But collective memory is constructed on unstable foundations. However far it may reach into the past, it conforms to present needs. Highly selective in the imagery it imports out of the past, it is easily bent as these needs change. Collective memory flourishes and weakens in accord with the vicissitudes of changing realities.

---

[55] Levi Smith, "Window or Mirror: The Vietnam Veterans Memorial and the Ambiguity of Remembrance," in *Symbolic Loss; The Ambiguity of Mourning and Memory at Century's End*, ed. by Peter Homans (Charlottesville: University Press of Virginia, 2000), 105–25.

[56] Benjamin Barber, *McWorld vs. Jihad* (New York: Ballantine, 2001).

The imaginative designs of collective memory operate in dialectical interplay with critical analysis, and can never withstand its subversions, at least in the pristine images in which they had first been called into being. That is why memory can never substitute for history based on solid evidence. The enthusiasm of collective memory cannot be sustained. At the crux of the dynamics of collective memory, though, is the notion of the eternal return. If memory is easily subverted, it resists forgetfulness. Its echoes continue to reverberate despite changing times and circumstances.

In a way, the historians' work in this time in which memory has surfaced in the realm of scholarship with such force and persistence suggests that we find ourselves at the end of a cycle of historiography. History is linear and privileges past and future; memory is cyclical and favors the present. Ironically, historiographical fashions tend to follow memory's cycle, for historical knowledge is not a simple aggregation of increasing information about the past. Topics of interest to historians emerge in light of present dilemmas, burgeon as they stimulate research, settle into narratives, lose their force in overspecialization, and in time are abandoned for new ones germane to the changing interests of a younger generation of scholars coming of age. Such a historiographical perspective draws attention to the topic under review in this study. Framing modern history as the saga of the building of the modern state, and illustrating it with the grandeur of its trial by war, no longer speaks to the needs of our times. In my view, the memory phenomenon in contemporary historical scholarship is a response to the dissolution of the realities that the ideologies of the modern age addressed. The preoccupation with memory in our times has permitted us to understand the imagination that inspired the commemorative projects of the modern era—what was valued in that era and how mourning was transfigured with the passage of time.

CHAPTER 4

Cultural Memory: From the Threshold of Literacy to the Digital Age

Orality/Literacy: A Brief Recapitulation of Early Scholarship

In ancient Greek mythology, Mnemosyne, the goddess of memory, was revered as the mother of the Muses of the arts and sciences. The ancient idea of memory was grounded in the concept of mimesis, which taught that memory and imagination are reverse sides of the creative act of "imitating nature." The tension between these two modes of understanding memory has endured ever since through all their reconfigurations. Tracing the history of the invention of new technologies of communication provides a key to an understanding of the changing relationship between memory's resources for preservation and those for creation, as new technologies advanced its staying power and localized its resources in memory banks external to the human mind. Though this approach to the study of collective memory was overshadowed in the 1980s by scholarship on commemorative and traumatic memory, interest in the modes of communicating collective memory has since become the most rapidly developing subfield of memory studies, and is today the focus of cutting-edge research on the dynamics of memory in our digital age. Certainly it was innovation and acceleration of electronic communication in the late twentieth century that spurred interest in like transitions in the past. Research in this field has since clustered around these times of transition between old and newly invented modes of communication: notably that from orality

© The Editor(s) (if applicable) and The Author(s) 2016
P.H. Hutton, *The Memory Phenomenon in Contemporary Historical Writing*, DOI 10.1057/978-1-137-49466-5_4

into manuscript literacy in antiquity (seventh century BCE → first century CE); from manuscript to print literacy in the early modern era (sixteenth → eighteenth century), and from print to media culture (late twentieth century). At each threshold, ideas about memory were reformulated. Each transition marked a significant departure from the modes of memory in cultures of primary orality.

The interest in media and memory as it emerged as a field of scholarship is indebted neither to Halbwachs nor to Freud, but rather to the Canadian scholar Marshall McLuhan (1911–1980). As prophet of the coming age of media, he sketched its significance at a time when the revolution in electronic communication was as yet only a dim horizon. His *Gutenberg Galaxy* (1962) was a highly original contribution to the nature of communication within print culture. He became a guru of the counter-culture during the 1960s, thanks to a little book, *The Medium is the Massage* (1967), one short on text and long on imagery. His sententious explanation of the phenomenon was the idea that "the medium is the message." It was to become the mantra of the communications revolution that would transform global culture in the late twentieth century. His brilliance notwithstanding, McLuhan could be obscure in his preference for evocative aphorism over transparent explanation.

More important for laying out the historical sequence of transitions in cultural communication was the work of his student Walter Ong (1912–2003). Ong is consistently lucid, and provides a panoramic view of the process of cultural communication as it evolved over time. Ong's work affirms his respect for the spoken word. He shows its staying power through all of the following transitions in technological innovation. Manuscript literacy continued to organize knowledge according to its protocols well into the modern era, that is, as topics rather than as linear indexing. He has much to say about the rise and democratization of print literacy in the early modern era, and he uses his discussion to dramatize the contrast between primary orality and print literacy. Ong, therefore, helped scholars to locate the deep sources of today's revolution in electronic communication in memory's foundational power. Before there was an electronic revolution there was a print revolution, and before that the long passage from orality into manuscript literacy deep in antiquity. As for the significance of the electronic revolution he characterizes it as a "secondary orality." But his work was too early to take the full measure of the cultural changes involved in this transition into our digital age. In Ong's interpretation, therefore, the history of the uses of memory correlates closely with inventions in the technologies of communication that issue from the primordial past.

Classicists had long been interested in the prodigious memories of the Homeric rhapsodes of ancient Greece. Early in the twentieth century, pioneering scholars Milman Parry and Harry Lord showed how storytellers in this milieu of primary orality relied on resources of memory largely abandoned today. Trained for the recitation of long epic poems, the Homeric rhapsodes displayed formidable powers of recall. These were enhanced by mnemonic techniques for stitching together episodes into a basic plot line with the help of formulaic phrasing. No one ever told the same tale in exactly the same way. Over centuries of oral recitation, moreover, these epics must have evolved imperceptibly with the changing realities of the times, for oral memory is a present-minded expression of a dynamic imagination. Parry and Lord buttressed their argument by observing Serbo-Croatian storytellers of their own day, who used the same mnemonic techniques and whose powers of recitation weakened dramatically once they were introduced to literacy.[1] The storytellers' uses of memory, they showed, are closely related to the technologies of communication available to them. The historical significance of Homeric studies to illustrate the cultural consequences of the transition from orality into literacy first appeared in Eric Havelock's *Preface to Plato* (1963), which traced the changing mindset of the Athenians from the Mycenaean (twelfth → tenth century BCE) to the Classical Age (sixth → fifth century BCE). Ideas expressed poetically in the speech of Homer were recast in a philosophical idiom in the writings of Plato, so that the meaning of the former was incomprehensible to the latter. In this way, Havelock explained why Plato came to believe that Homer "told lies about the Gods."[2]

The scope of such studies expanded and diversified during the 1960s. Originally of interest only to classicists and folklorists, the topic came to stimulate broad scholarly interest across the social sciences, thanks to the visibly expanding presence and ever more intrusive influence of media culture in the contemporary world. Scholars could see that the move from cultures of primary orality to those of manuscript literacy was but the first in a series of revolutions in the technologies of communication across two

---

[1] For an overview of the work of Parry, Lord, and other early students of orality/literacy, see John Miles Foley, *The Theory of Oral Composition* (Bloomington: Indiana University Press, 1988), 2–10.

[2] Eric Havelock, *Preface to Plato* (Cambridge: Harvard University Press, 1963) 3–15.

millennia that had transforming effects on perception, the uses of memory, and the organization of knowledge. In this heuristic perspective on technology as a force of change in the broad sweep of cultural history, scholars noted a long-range process of relocating reliable knowledge from the memory banks of a well-ordered mind into external archives available for public consultation. In each transition, the methods for organizing human knowledge were reinvented and the understanding of human memory reconceived.[3] The principal faculties of memory—imagination and preservation—originally so closely bound, over time came to be thought of as powers apart. Elements of this far-ranging approach to cultural history emerged piecemeal. An early pioneer was the Russian psychologist Alexander Luria, who during the 1930s conducted field studies of the effects of literacy on previously illiterate populations in Central Asia.[4] He noted rapid cognitive changes from a concrete to an abstract mindset with the advent of literacy. Historians of orality/literacy have learned much, too, from the fieldwork of anthropologists who have studied twentieth-century African communities in the midst of their passage from orality to literacy.[5]

From a historiographical perspective, much of the early work focused on the search for orality's survival within literate contexts. Among historians, the most practical advantage of this scholarship was a method for uncovering oral residues within written texts and so extending a historical reach into a realm of cultural memory otherwise inaccessible to the historian. Exemplary was the work of the Jesus Seminar, a gathering of biblical scholars brought together by Robert Funk and Roy Hoover, in search for the sayings of the historical Jesus. In their preface to *The Five Gospels; What Jesus Really Said* (1993), they set forth their method for interpolating the manuscript gospels to uncover the idealizations of the oral tradition of primitive Christianity. Jesus of Nazareth was a Jewish teacher speaking to a society in moral crisis, they argued, and his sermons were ethical,

---

[3] The changes in mentality have also been plotted by the anthropologist André Leroi-Gouhran as a five-stage process. These include oral transmission, written tables, file cards, mechanical writing, and electronic sequencing. See his *Le Geste et la parole: la mémoire et les rythmes* (Paris : Albin Michel, 1965), 65.

[4] Alexander Luria, *Cognitive Development: Its Cultural and Social Foundations* (Cambridge: Harvard University Press, 1976).

[5] Jack Goody, *The Interface between the Written and the Oral* (Cambridge: Cambridge University Press, 1987); Jan Vansina, *Oral Tradition as History* (Madison: University of Wisconsin Press, 1985).

not messianic. As a preacher, moreover, he never wrote anything down. The writings of the Evangelists, who exposed his life and thought, were composed some 40–60 years after his death. Disaggregating the mix in these texts of sayings attributed to Jesus, stories about him, and prophetic proclamations of his messianic role, they ordered these as an index to the sequence and roughly the dates in which the gospels were composed. Funk and Hoover convened a group of eminent research scholars to study the gospels as literary artifacts that encoded two generations of oral testimony. Participants in the seminar wrestled with these juxtapositions of first-hand testimony and later remembrance, seeking to factor out the pithy aphorisms that Jesus may have uttered from more elaborate idealization of his intentions in the oral tradition perpetuated by his followers.[6] For several years, seminar scholars debated their relationship by casting color-coded ballots for each passage of the major gospels. This sorting process became a basis not only for understanding the historical Jesus but also for fixing the dates of composition of these texts devoted to his memory on the basis of the degree to which they idealized his life and transformed his ethical sayings into theological prophecies.

The spread of print culture in the early modern era is the other major venue to which students of the technologies of communication gravitated, all the more significant because it signaled the crucial transition from ear to eye in the uses of memory. These studies opened a new perspective on the nature of the Enlightenment, shifting interest from the *philosophes* of this intellectual renaissance to the rapidly expanding cadre of readers eager to digest their teachings in a culture in which the printed word made knowledge more accessible to the public than ever before. Intellectual historians of an earlier generation once made much of the efficacy of the print revolution of the fifteenth century, for it was a factor in the success of the German Reformation. But today's students of the coming of print culture prefer its interpretation as a long revolution in the democratization of reading. Only by the eighteenth century was the subject matter of print culture sufficiently diversified and its public adequately literate to make manifest its far-reaching cultural effects. The Enlightenment, once studied for its writers, has for these scholars become as important for its readers, and it is out of their mindset that the modern uses of memory came to the

---

[6] Robert Funk and Roy Hoover, eds., *The Five Gospels; What Did Jesus Really Say?* (New York: Harper Collins, 1993), 1–38.

fore.[7] In the creation of a reading public, the Enlightenment witnessed the emergence of a "republic of letters" as a newly imagined community.[8]

The cultural effects of print literacy were evinced in two ways. On the one hand, printed matter moved facts to be remembered into books and encyclopedias, more accessible to far more people than had been the manuscript archives of an earlier age. In a subtle way, the active evocation of public memory through ready recall gave way to its private consultation in these compendia of knowledge.[9] On the other hand, the vastly expanded archival capacities of book culture freed the literate mind for a new kind of meditative introspection. The search for the self promoted a new reflectiveness about the resources of personal memory, which writers of the day portrayed as the deep source of personal identity. William Wordsworth and Jean-Jacques Rousseau inaugurated a Romantic cult of introspection in their autobiographical writings.[10] The modern novel, too, became a mirror for self-reflection, deepening the valuation of personal identity in the modern age. The novel as an aid to self-analysis received its most profound statement in Marcel Proust's *A la Recherche du temps perdu* (1913–27), which extolled the illuminating power of involuntary recall to transform the memory of a single incident into an entire milieu of remembrance.[11] The modern cult of private memory as soul-searching for personal identity would eventually acquire a scientific gloss in the psychoanalytic techniques of Sigmund Freud, who elevated the intuitive insight of the Romantics into a Positivist scientific principle.[12] This emerging divide between private

[7] See Elizabeth Eisenstein, *The Printing Press as an Agent of Change* (Cambridge: Cambridge University Press, 1979).

[8] Michael Warner, *The Letters of the Republic; Publication and the Public Sphere in Eighteenth-Century America* (Cambridge: Harvard University Press, 1990).

[9] Noteworthy is the contribution of historian Robert Darnton to the study of print culture. His early scholarship concerned the making of the *Encyclopédie* as the key tool for the organization and preservation of knowledge in the modern era of print culture. But his bestseller, *The Great Cat Massacre and Other Episodes in French History* (New York: Basic, 1984), reached a wider audience. In a series of artfully told stories, he canvassed the new social types born of the emerging age of print literacy: printers, hack writers, editors, clerks, and readers of novels. See also his *The Literary Underground of the Old Regime* (Cambridge: Harvard University Press, 1982).

[10] James Olney, *Memory and Narrative: The Weave of Life-Writing* (Chicago: University of Chicago Press, 1998).

[11] On involuntary memory in Proust's novel, see Daniel Schacter, *Searching for Memory; The Brain, the Mind, and the Past* (New York: Harper Collins, 1996), 26–28.

[12] Patrick Hutton, "Sigmund Freud and Maurice Halbwachs: The Problem of Memory in Historical Psychology," *The History Teacher* 27 (1994), 146–48.

and public memory reinforced the modern distinction between private and public life. Today's interest in the memory phenomenon has gravitated toward the latter. In terms of scholarly inquiry, individual and collective memories have gone their separate ways.

Jan and Aleida Assmann on Cultural Memory

German scholars Jan and Aleida Assmann redirected scholarship on collective memory onto a new pathway during the 1980s. Their interest turned to the way literate cultures build heritage. In this reorientation, they shifted from the process to the product, from the method to the content of such transmission. Whereas Havelock and Ong searched for remnants of oral phrasing that survived in literate texts, the Assmann investigate the long-term preservation of collective memory as it acquires material substance within the tangible domain of art, artifacts, images, script, and alphabets. They turned their attention from differences in the resources of orality vis-à-vis literacy to strategies employed across the ages for holding collective memory fast against the erosion of time. They label such content cultural memory. One might argue that their work was a response to the coming of the electronic revolution in the technologies of communication, which raised new issues about the stability of cultural remembrance. The study of cultural memory as a field of memory studies, therefore, dovetails chronologically with those dealing with commemoration and traumatic memory. The 1980s was the crucial decade in which these varied approaches took flight.

The studies by the Assmann follow Frances Yates in showing how the art of memory provided a significant pathway into cultural history. Yates's book on the art of memory, published in 1966, was the first to examine the intellectual uses of the art in its cultural contexts. There were two sides to Yates's approach. On the one hand, she offered a description of the method of the mnemonist. On the other hand, she explained how the classical art transcended its origins as technique of recall and came to be employed in the mnemonic schemes of Renaissance philosophers, who invoked memory's powers to interpret the workings of the cosmos.[13]

---

[13] Building upon oral protocols, Yates explained, they used the art of memory to provide a framework for building a body of cultural knowledge. She analyzed the uses of the art by neo-Platonic philosophers, with particular attention to their speculative purposes. Sixteenth-century magi, such as Giulio Camillo, Giordano Bruno, and Robert Fludd, believed that

The Assmann enlarged upon Yates's model.[14] They studied what humankind judged memorable in its cultural heritage and how over time it accumulated as a repository of human knowledge. Jan Assmann labeled this venture "mnemohistory," by which he means historical interpretation of the past not only for what it was but also for the way its memory was carried forward by posterity in a tradition of ongoing acts of remembrance. In this he identifies a middle ground between memory and history. The significance of a cultural memory resides in its capacity to inspire long-term remembrance. The reconstruction of such chains of memory is at once a study of communication following the method of Ong and Havelock and a study that expands upon the philological approach of Yates. The Assmann, therefore, address the cultural implications of new technologies of communication not as a historical succession of inventions but rather as an expanding body of knowledge. Their orientation is prospective rather than retrospective.[15]

One might argue that the line of inquiry pursued by the Assmann provides a counterpoint to Pierre Nora's French initiative. Whereas he and his colleagues in the *Lieux de mémoire* project deconstruct collective memory that has faded from view, the Assmann trace the elaboration of collective memory as it has contributed to the making of a memorable culture. Their particular interest is in the ways of memory's persistence in the variety of its art, architecture, and literature over long periods of time. Together they interpret how a memorable culture is stored, preserved, publicized, reworked, and displaced over time. The making of cultural memory, they argue, is the ongoing project of elaborating a heritage that will serve as an enduring record of what has been memorable in human creativity through the ages. To trace its history, however, is to confront the limits of this quest. Cultural memory evolves over time in its forms as in its meanings and so acquires a historical dimension in traditions of remembrance. The Assmann explain the way that heritage is transformed

---

their ornately decorated memory palaces mirrored the structure of the universe, and so contained the keys to its understanding. Celebrating the harmony between divine and human power of mind, their architectonic designs might be regarded as supernova of the intellectual quest of a waning philosophical idealism. Frances Yates, *The Art of Memory* (Chicago: University of Chicago Press, 1966), esp. 129–59, 368–72.

[14] Aleida Assmann, *Cultural Memory and Western Civilization; Arts of Memory* (Cambridge: Cambridge University Press, 2011), 18.

[15] Jan Assmann, *Moses the Egyptian; The Memory of Egypt in Western Monotheism* (Cambridge: Harvard University Press, 1997), 8–17, 21.

over time, sometimes fading into obscurity, at other times taking on new meaning as it is reconfigured in new cultural contexts.

The Assmann scan the making of cultural memory through significant stages in the making of Western civilization: Jan antiquity; Aleida modernity. Jan Assmann is an Egyptologist with an interest in the Mediterranean cultures of the ancient world—roughly the period 500 BCE to 500 CE (which, following Karl Jaspers, he refers to as the Axial Age). He was a leading figure of the Heidelberg group in memory studies that began work in the 1980s. Aleida Assmann is a literary critic, steeped in modern German and English literature and cultural history. They embarked on their scholarship only shortly after Pierre Nora's project had come to command scholarly attention during the late 1980s. For a long time limited to a German-language audience, their work over time captured the attention of scholars internationally, especially after it was translated into English about the turn of the twenty-first century. Their ideas now play out on a global stage. The key concepts with which they work are the canon and the archive.

Jan Assmann's Canon

Jan Assmann's work is significant for interpreting the passage from oral to manuscript literacy, for he opens a wider perspective on the cultural implications of the transition. He follows Maurice Halbwachs in his understanding of the nature of collective memory as a function of social power, but seeks to refine that concept by distinguishing communicative from cultural memory. By the former, he refers to living memory in the transactions of everyday life. Communicative memory is dynamic and reaches back no more than some 60–80 years—the time that episodic memories may be shared among living generations. Upon the death of each of these generations in turn, their contribution to communicative memory passes with them into the netherworld of time immemorial. By the latter, he alludes to the heritage through which the ancients sought to fix their collective identity in a more lasting way. It is this kind of memory to which he addresses particular attention, for it is a precondition of memory for the ages.[16]

---

[16] Jan Assmann, *Cultural Memory and Early Civilization: Writing, Remembrance, and Political Imagination* (Cambridge: Cambridge University Press, 2011), 6–11.

Assmann's discrimination between communicative and cultural memory reformulates Ong's distinction between orality and literacy, but in a way that sets aside the oral tradition that early scholars in the field found so fascinating. By contrast, he expands on the enduring material forms that cultural memory may assume. This kind of memory became possible only with the advent of literacy, and the art of memory was its first rudimentary framework. Cultural memory reflects the human quest for immortality in the commemoration of great deeds, events, and personalities, and as such provides symbolic foundations of group identity that transcend living memory. The initial cultural task had been to invent fixed points of remembrance with which to preserve continuity of identity between past and present and so to sustain the authority of a venerable past. To that end, cultural memory required the identification of salient objects that promote long-term remembrance.[17]

In this context Assmann introduces the concept of the canon. Today, he notes, we think of a canon as a masterpiece of literature. But it had a more practical definition in antiquity, for its original purpose was to provide a standard of measurement against which to evaluate new cultural creations and so to forestall innovation that might undermine foundational identity. The canon was the first step in establishing a fixed frame of reference in which to localize collective memory. A canon might be an architectural structure, an inscription, a text, or some combination of them. But in each guise, it served as a frame of reference with which to ensure ongoing cultural conformity. Based on a suspicion of innovation, the ancient canon revealed a profoundly conservative mindset. For the ancients, Assmann observes, the truth of heritage lay in its invariability. The canon served as a "counter-present," a bulwark of cultural stability to stay the tide of cultural innovation. Assmann interprets the canon in its origins as a foundational art of memory, providing an added dimension to Frances Yates's explanation of the ancient art as a rhetorical strategy.

Assmann explains that the understanding of the canon varied among the nations of the ancient eastern Mediterranean, as each fashioned its cultural identity in its own way. He distinguishes the nature and uses of canonization among Egyptians, Jews, and Greeks in the way each formed its cultural memories. His comparison highlights growing sophistication in the fashioning of cultural memory in a move from commemoration as rote

---

[17] Ibid., 36–44.

memorization to heritage appreciated in cultural contexts, way stations on a road toward historical understanding of tradition itself.

In its origins, Assmann argues, the canon was an Egyptian invention. The ancient Egyptians invested their cultural memory in the monumental architecture of their temples. These were literally memory palaces, symbolizing harmonizing correspondences between the earth and the heavens, the human and the divine. Over the course of two millennia, they built these structures according to the same precise specifications. Lists of their kings were inscribed on their columns. These lists were aggregative, neither revised nor interpreted in these updates. Some 700 names appeared on the columns of their early temples; 7000 on the later ones. The temple, therefore, served as the canon of ancient Egyptian cultural memory. The staying power of this practice of recording the dates of the royal succession was remarkable, commemorating continuity across 345 generations. This kind of commemoration, Assmann explains, never approached what we might characterize as historical thinking, for it allowed no room for interpretation.[18]

Among the ancient Hebrews, by contrast, the canon was rather a guide to righteous living. Whereas the cultural memory of the Egyptians was cosmological, that of the ancient Jews was ethical. Believing in a monotheistic deity with whom they claimed a special relationship, they looked to the past to understand God's plan for their destiny, and so interpreted their fortunes and misfortunes along the way as signs of his judgment on their ability to meet his expectations. Their understanding of their past was singular in that it was based not on its mythology but rather on its historicity. Myth recounts events that recur now and again. Historicity denotes events that happen but once and for all time.[19] In ascribing meaning to a particular history, Jews were unique among the people of antiquity.

The Pentateuch, the sacred books narrating the early history of the Jews, was their guide to understanding God's plan as a way to the redemption of their suffering. It recorded the history of their origins as a people, from the time of their exodus from Egypt under the leadership of Moses, through their wanderings for 40 years in the Sinai desert, to their entry into the promised land of Canaan. It recounted that experience as a time of divine revelation that set forth rules by which their descendants were commanded to live thenceforth. These texts, written by prophets, were

[18] Ibid., 87, 103, 156–65, 170–74.
[19] Ibid., 176–79, 230–31.

conceived as acts of historical remembrance. Moses, Assmann contends, was venerated not as a life and blood historical personality but rather as an iconic "memory figure" who personified leadership in the liberation of the Jews from their enslavement in Egypt. As an account of the past invested with sacred meaning, the Hebrew Bible became the Jewish canon, enshrining moral law for the ages. God's command to the Jewish people was to remember this past, explicit in its depiction as a saga of their historical journey. Historical remembrance became the mainstay not only of their identity but also of their hope for salvation.[20]

Assmann interprets Deuteronomy as the key text in the canonization of the Hebrew Bible, for it highlights the transition of historical remembrance among Jews from living to cultural memory. This sacred time in history aspired to fix Jewish identity for all time. While post-Exodus Jews would face new dilemmas, the meaning of that experience would be tested against references in the canonical narrative of the Hebrew Bible. As an extant text, it was open to ongoing commentary on its contemporary meaning, and so served as a "counter-present" with which to evaluate new situations. As immovable scripture, the Hebrew Bible launched a tradition of exegesis carried forward by rabbis, teachers respected for their learned interpretations of this formative period in Jewish history. In this way, the Hebrew Bible came to be contextualized within an ongoing tradition of interpretation, notably that of the Talmud, itself sealed as a canon by the fifth century CE. But the time in history worthy of interpretation remained constant through all the transformations in Jewish experience that followed. Assmann attributes the canonization of this time in Jewish history to the needs of the Jews in their Babylonian exile during the sixth century BCE, and in their subsequent subjugation first by Persians, later by Romans. Having no secure geographical place they might count as their permanent homeland, they internalized their identity in the recollection of their historic experience as recorded in their sacred writings. In obeying God's command for their remembrance, Jews formulated remarkably effective mnemonic techniques for perpetuating their identity as a people beyond that of any of their ancient contemporaries. They elaborated a liturgical calendar of rites and rituals designed to commemorate its most memorable events.[21]

---

[20] Ibid., 179–80.
[21] Ibid., 191–200.

Assmann presents the experience of the ancient Greeks as still another variant on cultural memory as it emerged in the ancient Mediterranean world. The Greek canon was more complex and its elaboration made critical historical thinking possible for the first time. Greek cultural memory came to reside within a canonical tradition of literature. Unlike the Jews, the Greeks had no sacred books, but rather an immemorial oral tradition of recitation of poetical epics that honored their Mycenaean ancestors in idealized myths of their heroic origins. The rhapsodes who told these epic tales were known collectively as Homer. In the seventh century BCE, the Homeric epics of oral tradition were set down in a newly invented alphabet borrowed from the Phoenicians. So flexible was this script that it facilitated the creation of other genre of literature, collectively contributing to the making of a pluralistic Greek cultural memory that flourished in the intellectual life of the city of Athens in the fifth century BCE. These works included written versions of the Homeric epics, the tragedies composed by dramatic poets, and the philosophical dialogues written by Plato in honor of his beloved teacher Socrates. Still recited more than read during those times, these texts were spoken at festivals and other celebratory events.[22]

The making of Greek cultural memory, Assmann explains, was furthered from the fourth century BCE onward by the assimilation of the Greek city-states into a cosmopolitan Hellenistic empire that mixed Greek with Persian, Egyptian, Syriac, Jewish, and other near eastern cultural traditions. In such a culture the meaning of the Greek heritage settled more deeply into literacy. Texts once intended to be recited now became texts to be read. The literary works of Greek culture became "classics" within a tradition of learned commentary on their meaning. So extensive became such writings that archives were constructed to house them, the most famous of these being the library at Alexandria. In these repositories, the classics were meant to be studied as a basis for placing them in interpretive context. Variations in their meaning became obvious in commentary written about them over time, opening the way to an unprecedented

---

[22] In his interpretation, Assmann qualifies Eric Havelock's thesis that the flourishing of the written word in the Greek Classical Age was largely a product of the transition to manuscript literacy made possible by the adoption of the Phoenician alphabet. Assmann argues that the politics of a changing society was also a contributing factor in the emergence of Greek cultural memory. From the seventh century BCE (the Archaic Age), the Greeks developed a new, more democratic politics suitable for the social mores of emerging urban city-states. As living memory, the way of life of the rural aristocracy had come to an end, surviving only as nostalgia for a mythic past of epic proportions. Ibid., 273–75.

historical awareness of the critical distance between the writing and the reading of a text, and so of the possibilities for ongoing interpretation. Historical remembrance, therefore, was a precondition for the rise of critical thinking among the Greeks. Assmann's depiction of this new mode of interpretation might be characterized as a rudimentary form of hermeneutical reading, that is, of making the classics familiar in otherwise unfamiliar settings. Drawing forth fresh meaning out of the classics to suit new situations developed into an intellectual tradition in which the cultural memory of the Greeks was perpetuated into modern times. Among these ancient traditions of cultural memory, Assmann maintains, only that of the Greeks enabled its literature to become the foundation of a canonical tradition of education that survives to this day.[23]

Aleida Assmann's Archive

In the last volume of his *Lieux de* mémoire, Pierre Nora delineated places of cultural memory that had contributed to the French heritage, and so served as a resource for his genealogy of the deep sources of the French national memory. From a methodological standpoint, his concluding survey of the places of memory of the French cultural heritage (which he refers to as *Les France*) resonates with the beginning of Aleida Assmann's study of the nature of cultural memory. Nora directed attention to the French cultural heritage retrospectively to uncover the cultural foundations of French identity at a time when French historiography as the pathway into its understanding appeared to have lost its way. Assmann takes the opposite tack, inquiring into the nature of cultural memory as it accumulated over the course of Western civilization. She reviews the many ways in which collective memory is stabilized in its cultural leavings, memorable artifacts that survive in a multitude of ways. Her work on the sustaining buildup of cultural memory might be construed as a reply to Nora's emphasis upon the crisis in postmodern historiography as a rationale for memory studies.

Taking as a point of departure Jan Assmann's distinction between canon and archive, Aleida Assmann shows how the concept of the canon in modern times has been transformed from a mnemonic standard of measurement into a memorable artifact open to ongoing interpretation. Its changing definition accompanied the decline of religion's original hold

---

[23] Ibid., 248–51, 262–71.

on its meaning. In modern times, the idea of a canon found its way into a broader realm of secular high culture. To put it differently, the canon as foundation of religious faith yielded place to the canon as "classic," valued for its claim to timeless esthetic appreciation. More specifically, the canon came to be identified as a place of memory in a tradition honoring the great literature and art of Western culture. The canon so reconceived continued to convey an aura of depth, seriousness, and thoughtfulness worthy of reflection.[24] Whereas Nora identified formative cultural memories in the French heritage randomly and ascribed to them no particular order of significance, Assmann sets forth a graded hierarchy of their importance as judged by posterity—from the most to the least memorable.

The canon, Assmann explains, is attended by archives for their interpretation. As repositories of neglected memories, archives provide a middle ground between living memory and passive forgetting. If canons serve as places of memory, archives provide the milieux that sustain them. They function as a "kind of lost and found department," she observes.[25] The contents of their holdings may in time be rediscovered, or recalled when conditions renew their relevance. They provide resource material for ongoing interpretation of the nature and meaning of heritage. As the canon expands, so too does the archive of commentary about it. Canons are texts; archives provide contexts for interpreting them, and as such may be consulted to disclose new meanings in different times and places. In light of such contextualization, the reception of a canon has a mnemohistory in the way in which it is remembered. The aura of the canon may continue to convey intimations of the sacred; the archive in the critical perspective it elicits leads toward profane understanding. In modern times, Assmann contends, the archive has come to rival the canon in importance, given rising attention to traditions of interpretation. Nor may the modern canon be considered timeless in the manner in which it had once been understood by the ancients. Canonical traditions are modified gradually, like an anchor being dragged with stubborn resistance from its moorings by the currents of the sea.[26]

[24] Aleida Assmann, "Canon and Archive," in *Cultural Memory Studies; An International and Interdisciplinary Handbook,* ed. Astrid Erll and Ansgar Nünning (Berlin: Walter de Gruyter, 2008), 100–02.

[25] Ibid., 106.

[26] Ibid., 98–104; idem, *Cultural Memory and Western Civilization,* 327–32.

In this respect, Assmann underscores the integral relationship between remembering and forgetting. Cultural memory requires forgetting, she explains, for memory is by its nature selective. Recollection disrupts the flow of experience, holding fast to a particular moment for reflection. Forgetting permits the memorable past to stand out in bold relief. But remembering and forgetting are not starkly opposed. She identifies two modes of forgetting: one active; the other passive. Active forgetting is tendentious, as in censorship or the destruction of documents and artifacts. Passive forgetting fades for want of immediate relevance or commemorative care. Much of her study concerns the hierarchy of significance attributed to human creations as they are remembered and remodeled over the course of time.[27] Over the long run, Assmann suggests, the sacred canon gradually concedes more significance to the profane archive. The archive is oriented toward the future, betting on the assumption that a classic will withstand the ravages of time by virtue of its ongoing capacity to stimulate commentary. The archive may house the past, but is beholden to the future. As she explains: "The archive is the basis of what can be said in the future about the present when it will have become the past."[28]

As a practical matter, Assmann allows, archives have tended to be constructed by statesmen, only to be appreciated later by historians. Governments maintain archives to house records that lend authority to their power. But political interests are likely to change rapidly, whereas the archives they engender prove more enduring. Over time, the archive, originally constructed to reinforce a framework of political interpretation, eventually loses touch with it. Old records acquire greater interest among historians in their efforts to place politics, conceived in the immediacy of its everyday practices, in an extended temporal context. Assmann therefore notes a curious paradox: the modern idea of progress is coeval with the rise of a new antiquarianism. As she puts it, the "first life" of the archive is displaced by a "second life," open to a far larger, more politically neutral kind of interpretation.[29]

One might contend that Assmann's interpretation of cultural memory provides her version of a memory palace, or better, a memory pyramid.

---

[27] In making her case about the nature of cultural elaboration and transmission, Assmann draws upon the insights of the largely forgotten German art collector Aby Warburg, *Cultural Memory and Western Civilization*, 163, 198, 214–16.

[28] Assmann, "Canon and Archive", 102.

[29] Ibid., 103.

She formulates a sliding scale of the significance of memorable artifacts, as judged by posterity. At the apex the works of the canon are enthroned. They are recognized in a sustained way over time as the greatest intellectual and aesthetic achievements of civilization. In the strata below may be found consciously constructed archives that house documents and other memorabilia of potential interest. They are too precious to be discarded. As data carriers they assume varied forms, from simple boxes to enormous buildings. The data they contain may be consulted easily and periodically recalled, providing in their ensemble a "metamemory" to which living memory has recourse.[30] Outside the archive, lesser leavings of cultural memory may be located on the downward slope of the pyramid. Attention may fall on these neglected traces of the cultures of the past, recovered from time to time as archaeological finds. Beneath these lie the waste products of human creation, which, though discarded, remain residues that may occasionally be rescued to enhance human knowledge of the past. Rarely if ever, Assmann concludes, does a cultural object disappear into an oblivion that renders impossible the recovery of its meaning. The framework that Assmann devises lends nuance to the long-standing distinction between high and popular culture. In cultural memory, the greatest cultural achievements are given pride of place. But all the remains of culture find their niche in a hierarchy of assessed value. Canons receive our constant attention; the holdings of archives may be recalled; traces of culture may be rediscovered; waste may be rescued.[31]

What is distinctive about Assmann's pyramid of cultural memory is its materiality. The material artifact, she explains, was perceived to be a guarantor of its stability. But even in material form, the pyramid is no static edifice. Its building blocks may be rearranged. Like tectonic plates, they move with the times. Considered over time the pyramid of cultural memory is a work in progress. In its extant arrangements, nothing is remembered in its present format forever. Cultural memory is useful only so long as it provides a mirror for the mind as a record of its creations. Assmann suggests that this severing of knowledge from the intention that inspired it—a divide between knower and known—harks back to Plato's discussion of the effects of writing upon memory in his Socratic dialogue *Phaedrus*. Writing, Socrates explains, introduces a different kind of memory, one based on representation rather than experience. The living memory of

---

[30] Assmann, *Cultural Memory and Western Civilization*, 106, 196–97.
[31] Ibid., 201, 369–76.

communication gives way to the referent memory for posterity. For Plato, Assmann argues, this passage from living to cultural memory reveals the "tragedy of culture," leaving dynamic living memory behind in favor of fixed forms of cultural remembrance. However memorable, these forms soon acquire critical distance from the experience they were designed to commemorate.[32]

Herein lies the key to Assmann's foundational interpretation of the relationship between living and cultural memory. In its elusive ambiguity, memory is at once steadfast and fickle, for it has two faces—one as art (*ars*) and one as power (*vis*). These are its preservationist and creative modes, and they have a dynamic relationship. The arts of memory aspire to create stable structures in which to hold fast a memorable past. But memory in its resources for creativity resists the constraining tug of preservation. Its impulse is to move on, and it continually aspires to reshape the past in new constructs that reflect the realities of the present. Living memory, therefore, constantly reconfigures the past. The products it commemorates are eroded by the process in which it is dialectically engaged. Assmann therefore pays close attention to the dynamic relationship between the active mind in the present (memory as *vis*) and remembrance of its past creations (memory as *ars*). She devotes most of her attention to the latter, *ars* writ large in the varieties of its cultural representations. But memory as *vis* is always present in her analysis. The creative power of mind finds fulfillment in the products of its creation. Such power, moreover, is inspired anew by reflections on them as they are called to mind as the heritage of cultural memory. Her interpretation, therefore, turns on the reciprocal relationship between the two. Just as the canon is the highest expression of *ars*, so sublime recollection (anamnesis) is the ultimate expression of *vis*. Their interplay reveals the way in which the human mind remains aware of its creations. Cultural memory as the archive of human achievement is the resource that triggers the creative power of the human imagination. The archive serves as a frame of reference and as a resource on which living memory relies to maintain its present identity between past and future. Human needs are redefined in response to changing realities, and the value ascribed to heritage with it. As the mind turns to new projects, so it will view the archive in a different way.[33]

---

[32] Assmann, "Canon and Archive," 104.
[33] Assmann, *Cultural Memory and Western Civilization*, 17–22, 98–99.

Considered from the perspective of our times, Assmann concludes, the divide between memory as *ars* and memory as *vis* grows wider, a precondition for the crisis of memory we face as archival records grow exponentially, particularly with the advent of the digital age. Even as a cultural force, the living memory of oral tradition relied upon ritual performance, the sameness of repetition again and again. Cultural memory, by contrast, depends upon particular representations open to ongoing interpretation. Accordingly, Assmann explains, the long-range historical trend has been the expanding power of new technologies of communication to export cultural memory into an ever more complex array of archives for housing human knowledge. The print revolution was a major step in this process. It ushered in what might be regarded as a golden age of cultural memory. Print culture encouraged profound reflection in plumbing the depths of the archives of cultural inheritance. Accordingly, it marked the ascent of historiography as a mode of cultural remembrance. Historians then boasted of living in the archives, assiduous in their pursuit of evidence of the way humankind has constructed its cultural world. Their task, they argued, was to publicize the findings they had discovered hidden therein.

For Assmann, this harmonious interaction of text and context has changed with the coming of the digital age. The transition is at once the story of the dissolution of old archival frameworks coupled with the seemingly limitless resources of the archive reconceived as an electronic medium. The electronic archive has made possible the exponential expansion of its holdings while easing its accessibility. Its storage capacity in cyberspace is boundless. Ironically, it has led to the mobilization of mass data, and imperceptibly prepared the way for the archive to take on a life of its own. Hitherto, Assmann explains, the essential nature of the archive resided in its materiality—books, documents, artifacts, and lesser memorabilia. But in the digital age, its material holdings are being exported at an accelerating pace into the immateriality of cyberspace. Its formats have been opened to reconfiguration in the process. The book in its tactile materiality, Assmann speculates, threatens to become an obsolete artifact. Artists are already depicting it in the guise of a fading art of memory.[34]

In the transition from materiality to immateriality, the meaning and function of the archive is being revised. In the digital age, Assmann contends, the archive is in the process of a transformation into an immaterial repertoire of data carriers whose cultural consequences we have yet to fully

---

[34] Ibid., 200–01, 340, 344–57.

fathom. During the age of print culture, the archive had been a resource for enhancing the powers of mind in its quest for deeper understanding. Its holdings enabled researchers to plumb its depths in the expectation that there was wisdom in the past that would nurture understanding of the present. The great benefit of the material archive as a repository of tangible artifacts was the stable record it contained, awaiting recall. The electronic archive, by contrast, fosters shallow if efficient information retrieval. To put her point differently, the pursuit of knowledge recedes in the face of the gathering of information. Today the archive is mined for data that may illustrate the tendentious needs of political and economic power, shedding all pretense of its once valued function of illuminating the autonomy of the past vis-à-vis the present.[35]

The archive of the digital age thereby erodes the boundaries that once defined the nature and limits of cultural memory. The effect has been to strain the human capacity to maintain the relationship between living and cultural memory. In these circumstances, the digital archive threatens to take on a life of its own, as it is transported into cyberspace and sheds its materiality. In surveying long-term remodeling of the archive in the move from print to digital formats, the archive is being reconceived as an active resource of artificial intelligence. The archive of the future, she speculates, may begin to mimic the operations of the human brain. Under these circumstances, the bond between *vis* and *ars* will loosen, as the archive reorganizes and remodels its holdings on an ongoing basis. Whether *vis* will be able to master *ars* is an open question. Already the fulcrum of power shifts in the direction of the archive's autonomy. In the process, Assmann worries, memory as *vis* and memory as *ars* lose contact with one another as they go their separate ways. The cultural inheritance of the past as resource and stabilizing reference loses its bearings. As cultural memory surrenders its mirroring power to the processes of information technology, anxiety over the cultural world that we are leaving behind grows more acute. Cultural memory is the foundation of our collective identity. How it will serve that end in the digital age awaits our ingenuity and unsettles our expectations.

[35] Ibid., 340–41.

MEMORY STUDIES IN THE DIGITAL AGE

"Memory Unbound"

In dealing with memory in an age of electronic communication, we enter a new cultural sphere, a realm for memory at once more expansive and diffuse than understood before.[36] Some scholars identify it as a "third wave" of memory studies that address the nature of collective memory in terms of a diverse array of new formulations.[37] With an accent on cultural memory in our digital age, they base their analysis on a number of new historical phenomena: a rapidly expanding array of new technologies of communication, a globalizing network of communication, and perhaps most importantly for historians, an illusory perception of the accelerating pace of time in the contemporary age. Broadly speaking, the possibilities of encoding cultural memory in digital formats has furthered the archivists' ambitions to expand the preservation of the past, for its record is so easily incorporated into digital archives and so readily mobilized for present use. At the same time, it has lost its tangible fixedness as a referent out of the past. In an age of digital technology, the remembered past moves with the times. As media change, our understanding of the nature and uses of cultural memory change with it. From a historical perspective, memory's preservationist role has yielded place to its creative capacity to incite the imagination, especially among students of communications theory today.

Since the turn of the twenty-first century, the topic of memory has visibly become an ever more integrated interdisciplinary interest, and it plays out on a global plane. One speaks today of "memory studies" rather than of memory and history. Work in this field is now typically collaborative, its finding more often published in anthologies of articles rather than in books by individual scholars. Scholarship has moved from an interest in the historical modification of the way static objects are remembered (famous person-

---

[36] The term "memory unbound" has been widely used by scholars to denote the turn toward the interest in digital memory.

[37] The idea of a "third wave" of memory studies is common currency among the generation of scholars coming of age today. See, for example, Gregor Feindt, Félix Krawatzek, Daniela Mehler, Friedemann Pestel, and Rieke Trimçev, "Entangled Memory: Toward a Third Wave in Memory Studies," *History and Theory* 53 (February 2014), 24–44. They focus on the "entanglement" of diverse approaches to collective memory, and repudiate Nora's regret over the loss of unity and homogeneity in national memory. They explore the possibilities of what "European memory" might be.

alities and events) to an appreciation of the dynamic, incessant movement of cultural memory, as the past is mobilized in the present in ever changing ways. As scholars in "memory studies" profess, the earlier approach focused on memorable objects out of the past to which posterity wished to hold fast. This was inevitably a losing battle, as memories faded for want of recognition, were contested as reconsidered in the context of new political and cultural contexts, or were idealized in abstract ways that lost touch with the concrete historical realities they wanted to evoke. Today's scholarship emphasizes memory's mobility, its images used and reused in a myriad of formulations. This approach advances the idea of collective memory as a resource for rethinking present-day culture in light of the dizzying pace of change among newly invented modes of communication. In a fast-moving world of technological innovation, human memory has begun to mimic the technologies through which it is communicated. The key to its appeal is the speed of communication, which approaches the instantaneous. In such circumstances, new devices become obsolete in ever shorter time spans.

Cutting-edge research in this field, moreover, places issues of cultural memory in a global rather than a national context, moving away from the kind of scholarship on national commemoration in which the field began. Media culture has reshaped collective memory—not only in the powerful ways in which it archives and mobilizes mnemic images but also for the way in which it integrates images drawn from earlier advances in the technologies of communication. It incorporates the full resources of photography, film, video animation, television clips, as well as the older technologies of print culture. Unbound from heavy reliance upon print formatting for its evocation, the past is now being imported into the present in these multifold and varied ways.

If Nora's framework for collective memory resembles a memory palace, and that of the Assmann a memory pyramid, literary scholars Astrid Erll and Ann Rigney present a platform in what might be likened to a memory theater. In their recently published anthology, *Mediation, Remediation, and the Dynamics of Cultural Memory* (2012), they, together with their contributors, set out to explore the impact of digital media upon cultural memory. Their work draws heavily on electronic age communication science scholarship, and has inspired the idea of a "third wave" of memory studies.[38] Their work involves a reformulation of the idea of cultural mem-

---

[38] Astrid Erll and Ann Rigney *Mediation, Remediation, and the Dynamics of Cultural Memory* (Berlin: De Gruyter, 2012).

ory as advanced by Jan and Aleida Assmann, who put their accent on its material nature. Memory reconceived for the digital age, by contrast, appears as a stage on which the past is reenacted in the present. As they explain, today's work on cultural memory aspires to integrate intangible as well as tangible cultural artifacts of all sorts into a reconceived theory of communication—"spoken languages, letters, books, photos, films, computers, communication devices," all of which in their varieties contribute to the remaking of cultural memory in cyberspace. Students of cultural memory no longer formulate transitions from one mode of transmission to another as sequential (as in, for example, the transition from orality into literacy). All genres of media are recognized as coeval in their capacity for mobilization. To express the notion succinctly, media are synchronic and synergistic (i.e., vibrating in ways that stimulate interaction).[39]

Erll and Rigney also challenge the proposition that cultural memory strives for commemorative consensus, an earlier conception which tends to get lost in nostalgia and is based on a deterministic model of entropy. Rather, they argue, cultural memory is paradoxically revitalized by contestation, to the degree that it seems worthy of interrogation as a way of understanding the meaning of the past. That is why these scholars put the accent on the "dynamics" of memory. In this respect, they avoid the term "progressive" in characterizing modes of communication, which connotes a linear conception of time. In reformulating the idea of cultural memory in this way, Erll and Rigney arrive at an assessment reminiscent of explanations of the dynamics of primary orality. Now as then, there is no turning back to some foundational event fixed in the past as a point of origin, but rather the incessant updating of the past as it is imported into the present. They summarize the rethinking of the memory question this way: Earlier interest sought to localize memorable artifacts as reference points securely fixed in the past. Recent interest, by contrast, shows how memorable images out of the past recirculate in present milieux of memory. Cultural memory in the digital age is protean and continually on the move. In the context of digital technology, the relationship between past memory and present perception is reconceived as emergent rather than as retrospective. One uplifts the memorable past into present conceptualization rather than harks back to its place in linear time.

[39] Astrid Erll and Ann Rigney, "Introduction: Cultural Memory and its Dynamics," in ibid, 1–11.

Erll and Rigney point to the significance of the pioneering study by Jay David Bolter and Richard Grusin, *Remediation* (1999), who disassemble the process of incorporating past and present into three memory cycles: remediation; hypermediation; premediation.[40] Briefly, Bolter and Grusin explain the ordering of memory in the digital age as follows:

Remediation concerns the refashioning of cultural memory to suit the needs of the present moment. Cultural memory deals with processes not products. In its digital guise, it is dynamic not static; it works through updating rather than continuity. New media refashions old media as they employ them in a wider context. So it is not just cultural memory that is refashioned but media itself. Remediation, therefore, has a double logic. It at once strives for immediacy (presence) and hypermediacy (critical distance), which are joined in perpetual oscillation.

Hypermediation is the awareness of the varied media through which cultural memory may be simultaneously deployed. In the digital age, cultural communication comes in a spatial design, a constant reminder of the multiplicity of modes of evoking the past. The concept of hypermediation revisits the idea that no extant medium is ever discarded. Rather, it is incorporated into a repertoire of new and more advanced technologies with which it continues to interact. The four stages in the invention of media—from primary orality to manuscript literacy to print culture to digital media—are utilized synchronically, not as a sequence of stages along the way. A hypermedium, such as a computer screen, may display the full array of its forms simultaneously—speech, photos, film clips, video animation, live televised images. The effect is to juxtapose immediacy and critical distance. The capacity of digital media to communicate images approaches the instantaneous. The display of varied images on a single screen reveals the multiple perspectives in which cultural memory may be appreciated at a glance. The watchword of media is the acceleration of time, and computer scientists strive to close that gap between past and present. One might characterize the presence of the past in such presentation as a syncretism of mnemonic display. Hypermediation provides context in the use of digital technology. Its power lies in the variety of perspectives it opens to the viewer. Or to put it in the context of our study, hypermediation is to remediation as historiography is to history.

---

[40] Ibid., 3–5; Jay David Bolter and Richard Grusin, *Remediation; Understanding New Media* (Cambridge, MA: The MIT Press, 2000), esp. 20–62.

From remediation in the present, there is the possibility of "premediation," an infinite regress toward earlier formulations of the same image. As Erll and Rigney contend: "No memorial monument is thinkable without earlier acts of mediation."[41] Premediation involves a reaching back in search of earlier ways in which cultural memory had been cycled and recycled. The concept evokes the idea of infinite regress, in the manner made famous by Michel Foucault and Carlo Ginzburg. In looking back in remembrance, there is no fixed beginning on which one may cast an anchor. For all commemorative images mimic earlier ones, descending into time immemorial. It is worth noting, too, the degree to which studies of remediation bear so many of the hallmarks identified by Ong as characteristic of memory in cultures of primary orality: dynamic memory, agon, updating, recycling, presence of the past, memory in its performative mode. Ong's observation on the electronic age as a "secondary orality" may have seemed gnomic at the time he wrote about it. Here in digital format, however, cultural memory becomes a theater that displays the past for present consumption. The orientation promoted by the media revolution resonates with presentism as a conception of historical time—what the French historiographer François Hartog has referred to as today's "regime of historical time," one that privileges the present moment over the perspectives of past and future as the essential frame of reference for historical understanding.

As communications scholar and contributor to this project Andrew Hoskins points out, the archive upon which digital memory draws is infinitely expandable into "sublime amounts of information." Yet digital memory is highly unstable. In such circumstances, the very idea of the archive is reconceived, for its potential holdings are boundless. The idea of memory as performance is linked to corollary notions of "platforms" and "theaters," places where the presence of the past comes alive. The key point is that remediation amalgamates two of the historiographical strands that had developed along the way: that of commemoration and that of orality/literacy, and takes them in a new direction. Digital communication, Hoskins argues, obliges us to rethink the nature of the archive. Once understood as a place of storage of cultural memory, it has come to be conceived as a network of dynamic processes. Digital communication erodes

---

[41] Erll and Rigney, "Introduction: Cultural Memory and its Dynamics," 4. See also Erll, " Literature, Film, and the Mediality of Cultural Memory," in *Cultural Memory Studies*, 392–95.

a sense of material placement. Here he wishes to show the limits of Aleida Assmann's conception of cultural memory. For her in her studies of the modern era, cultural memory is encoded in its material memorabilia. But digital memory is immaterial, and hence operates according to different principles. The past is no longer conceived as "punctual" (i.e., localized in a particular time and place), but rather reconceived as an emerging network of communication. In the digital archive, the past is continually present, engaged in an interactive process in which the old distinction between past and present is superseded by the conflation of past and present into a synchronized present. That is why students of digital memory argue that all communication is mediated. The process never pauses, let alone stops. The paradox of digital communication is that it is at once readily available yet easily obliterated. Traces of the past are vulnerable to being expunged in their recycling (i.e., remediation).[42]

Memory studies in our globalizing digital age mark a reorientation in the conception of the relationship between past and present. The study of memory has escaped its niche in historiography, as the idea of cultural memory has been reconceived in order to place an accent on memory as dynamic and synchronic in opposition to earlier studies in history, which were perceived to focus on the fixed and the diachronic. In this conception of memory in the digital age, the past is recycled for new uses rather than merely preserved for intellectual edification. This reorientation repudiates the idea of a fixed point of departure in favor of a genealogy of infinite regress. This innovative approach, nonetheless, conveys intimations of the venerable idea of the eternal return. The archive of memory as a concrete repository of traces of the past has been transported into cyberspace where it sheds its materiality. It reaffirms our understanding of the way old technologies are nested within the new. No old technology has been abandoned, but rather is redeployed in new ways. It relativizes the past in the quest to serve present need. The old emphasis on the continuity of memory in the dynamics of tradition give way to "updating" the past as it is integrated into present understanding. A past once revisited for static objects that invite reflection now yields to a past actively mined for memories that might be mobilized in newly creative ways in the present. In this conception, old memories live once more, revitalized in new cultural contexts, reformatted in the digital technologies of the contemporary

---

[42] Andrew Hoskins, "Digital Network Memory," in Erll and Rigney, eds., *Mediation, Remediation, and the Dynamics of Cultural Memory*, 91–101.

age. Drawing on pioneering work by communication scientists Bolter and Grusin, students of digital age memory reaffirm the notion that the medium of communication is not a mere conveyer of information, but also a means of shaping cultural memory.

This approach to memory would have unsettling effects for historians as it cut itself free from its roots in material culture in the late twentieth century to become reformulated in the emerging field of memory studies. History as it had developed as a profession was bound to the material archive. It provided the place of historical research and the professional home of historians. It was the place where they worked and in which their imaginations came alive. In effect, the archive was the historians' place of memory. It was there that they were initiated into their calling, and where the most serious among them would thenceforth dwell. For historians, the archive defined their sense of calling, and shaped their understanding of the past in relation to the present.[43] The existential past was fixed at a remove in time. The historian endeavored to understand it and to represent it for the present, but without the belief that its existential realities might be exhumed to live again. In such representations, one might evoke authentic perspectives on the past, but never recapture it completely. The past, historians believed, could not be resurrected. Rather the historians' task was to widen the horizons of present understanding by showing how different the past was from the present and how the distance between the two was traversed in a process of ongoing change. In this exercise, the historian would reveal the pathway we have traveled through time toward our present circumstances.

The notion that the past could be imported into the present, skipping understanding of all the stages along the way, was unnerving for historians in light of their training and expectations, for it destabilized what they had understood to be the object of their investigations—the evidence securely fixed in the holdings of the archive and the printed books drawn from their resources. But in an age of media, the once immovable foundations of the archive began to dissolve in their transformation into immaterial repositories. The past would be preserved; indeed, more of its traces could be housed and retrieved than ever before. Its digital capacity for collection and retrieval of data was virtually limitless. Yet the framing

---

[43] See Carolyn Steedman, *Dust: The Archive and Cultural History* (New Brunswick, NJ: Rutgers University Press, 2002), 66–88; and the review by Jo Tollebeek, "'Turn'd to Dust and Tears': Revisiting the Archive," *History and Theory* 43 (May 2004): 237–48.

of such knowledge had become provisional, and its holdings subject to imperceptible remodeling in the process of format updating. The archive was more encompassing than ever in its exponential expansion. Yet it had sacrificed its claim to permanence in fixing the representation of the past. For the present age, the idea of memory is on the move.

"History Entangled"

Meanwhile memory studies as an interdisciplinary venture abound, linking American and European scholarship in an expanding network of intellectual exchange. As scholars from across the curriculum explore the myriad of ways in which cultural memory may be conceptualized, they are ever more inventive in their formulations, including such notions as "entangled memory," "palimpsestic memory," "digital memory ecology," and "prosthetic memory." Their frameworks suggest the global reach of collective memory reconceived in various notions about transcendence: "transnational," "transgenerational," "transmedia," and "transdisciplinary."[44] At the same time, these rubrics are applied in precise, sometimes obscure case studies chosen from around the world. The tenor of these studies is mildly political in championing the cause of human rights. Alison Landsberg's notion of "prosthetic memory" is exemplary. It suggests broader, more humane sympathies for strangers, beyond personal experience, or allegiance to family or nation.[45] Herein, memory and history, far from being interpreted as oppositional as they were in an earlier age, are thought of as being "entangled." Landsberg lays emphasis on the way mass media, notably film and television, have made it possible for us to identify vicariously and with deep emotion the experience of others with whom we have no personal or social connection. So reconceived, historical studies of memory remain important, but now under the aegis of mnemonic conceptual frameworks. Studies of major historical figures and events out of the past have an important role in this avenue of research, but are of interest as memorable symbols embedded in cultural legacies or as "mnemonic signifiers," shorthand for historical forces in play in today's politics.

[44] Feindt, "Entangled Memory"; Lucy Bond, Stef Craps, and Pieter Vermeulen, eds., *Memory Unbound; Tracing the Dynamics of Memory Studies* (New York: Berghahn Books, 2016).

[45] Alison Landsberg, *Prosthetic Memory; the Transformation of American Remembrance in the Age of Mass Culture* (New York: Columbia University Press, 2004), 19–21.

CHAPTER 5

From History to Historical Remembrance in Holocaust Studies

TRAUMA AND MEMORY: THE HISTORIANS' DISPUTE

The third and for a time the most intensely studied approach to memory studies concerned the historical task of recovering memories of the Holocaust, both personal and collective. Germany rather than France served as the setting for this historiographical redirection. During the 1980s, the historians' reckoning with the Holocaust took center stage, and the consideration of its place in German history turned what had once been conceived as a metanarrative of nation-building toward the future into a fragmented tale of how the present age might redeem the past for its errant ways. Over the course of the decades following World War II, the genocide of European Jews by the Nazis took on greater moment among historians as an "unmasterable past" that demands careful reflection as a prelude to writing its history. In light of the intensity of the controversy about "historicizing" the suffering of the victims of the Holocaust, the relationship between trauma and memory came for a time to overshadow other approaches to the question of memory's relationship to history. This effort among German scholars to come to terms with this shameful episode was dramatized in the "Historians' Dispute" of the mid-1980s over the conditions under which the memory of the Holocaust might be permitted to pass

© The Editor(s) (if applicable) and The Author(s) 2016
P.H. Hutton, *The Memory Phenomenon in Contemporary Historical Writing*, DOI 10.1057/978-1-137-49466-5_5

into history.[1] There were questions about how that transition might be accomplished in the face of the still untold stories of victims of Nazi persecution demanding public recognition and accountability before they might pass into any settled account of Germany's past. Then, some scholars contended, that narrative would have to be rethought and rewritten. Meanwhile, they argued, historical judgments about the meaning of the suffering of the victims of Nazi crimes should be held in abeyance, leaving a void in the story of modern German history not to be filled until all repressed memories of its horrors had been uncovered and acknowledged. For some scholars, the controversy remains unsettled to this day.

The "Historians' Dispute" (*Historikerstreit*) in West Germany dramatized an argument about the relationship between memory and history in coming to terms with the Holocaust. Ernst Nolte, a scholar well-known for his studies of fascism, proposed that 40 years after the episode, it was time to "historicize" its memory. A number of scholars challenged his proposal, contending that any such interpretative assessment was premature. Frankfurt School scholar Jürgen Habermas, for example, argued that the narrative of German history, derived from the imperial ambitions of the Wilhelmine era, was itself in need of thorough reconceptualization.[2] It was not yet time to locate the Holocaust within any narrative context. Too many unresolved moral issues still needed processing. The controversy raised the question of whether the meaning of the Holocaust could ever be adequately treated through historical interpretation, given the exceptional nature of its atrocities. Following Habermas, some scholars pondered how historians could adequately convey the suffering that victims knew. Here was a realm of memory that seemingly defied historical interpretation.[3] In this way, the Historians' Dispute of the 1980s led into a debate about the limits of representation during the 1990s.

Soul searching about traumatic memory as a legacy of the Holocaust was not limited to German scholars. The French historian Henry Rousso

---

[1] Charles Maier, *The Unmasterable Past; History, Holocaust, and German National Identity* (Cambridge: Harvard University Press, 1988).

[2] For the debates themselves, see the collections by Peter Baldwin, ed., *Reworking the Past: Hitler, the Holocaust and the Historians' Debate* (Boston: Beacon Press, 1990); James Knowlton and Truett Cates, eds., *Forever in the Shadow of Hitler? Original Documents of the Historikerstreit* (Atlantic Highlands, NJ: Humanities Press, 1993).

[3] Saul Friedländer, ed., Probing *the Limits of Representation; Nazism and the Final Solution* (Cambridge: Harvard University Press, 1992), 1–21.

offered a parallel study concerning French complicity in the Nazi project. In some ways, the French case was more difficult to confront, for the French in the postwar era had been more reluctant than the Germans to admit their role in speeding Jews to Nazi concentration camps. But Michael Marrus and Robert Paxton's well-documented study of the plight of Jews in Vichy France, published in 1981, obliged French historians, and the French public, to take a hard look at their morally compromised past.[4] Their book raised issues that led belatedly to some sensational prosecutions, and exposed the way in which Vichy's cooperation with the Nazi vitiated the careers of some of France's eminent politicians, even the highly regarded socialist president François Mitterrand.[5] It was for this reason that Rousso's *Vichy Syndrome* (1989) had such a profound effect on French studies of the war years. Like the German scholars, he cast his argument in psychoanalytic terms. He argued that after the war French leaders put away their unhappy memories in the name of starting over under the banner of national reconciliation. Some apologized for Philippe Pétain's pliant collaboration as a necessary expedient for a defeated nation, and all agreed that it was better to move on. The repressed memories of unresolved issues relating to the war, however, continually resurfaced in its aftermath with undiminished vehemence. In each postwar crisis, and most ardently in that over the future of Algeria, the "Vichy Syndrome" stirred up unresolved political antagonisms. As Rousso and his co-author Eric Conan entitled a follow-up book on the subject, this was "a past that would not pass away," a troubling memory that defied historical evaluation that would put it to rest.[6]

Scholars agree that the Holocaust was an exceptional event, unprecedented and of a magnitude of infamy impossible to match. But cultural historian Alon Confino has recently challenged the notion that its history remains an "unmasterable past." As a historical episode, he explains, it resides on the extremes of human suffering, but from a historical standpoint it is no more inscrutable than any other historical event. He argues that the discussion of the Holocaust in history is overdetermined in its recourse to psychoanalytic vocabulary and to its exaggerated claims about

---

[4] Michael Marrus and Robert Paxton, *Vichy France and the Jews* (New York: Basic, 1981).

[5] Pierre Péan, *Une Jeunesse française; François Mitterrand, 1934–1947* (Paris : Fayard, 1994), 202–27, 317–25.

[6] Eric Conan and Henry Rousso, *Vichy, un passé qui ne passe pas* (Paris: Gallimard, 1996). See also Joan B. Wolf, *Harnessing the Holocaust; The Politics of Memory in France* (Stanford: Stanford University Press, 2004).

repressed memory. We do not master the past by arriving at a consensus that brings "closure," that is, settles all problems for all time. Rather, we learn from the experience of the past to which we may never be completely reconciled, and each generation must do so anew. He wonders whether Holocaust memory was ever as completely repressed during the immediate decades after the war as some scholars have argued, and suggests instead that it is rather a growing awareness of the historical magnitude of the evil of the Nazi project that has become more evident over time. Certainly, there was more public discussion of the Holocaust during the 1970s and the 1980s than there had been during the postwar decades, but that may have to do with the way its remembrance was publicized over time in a widening array of films, books, and other exposés that made the public aware of its distinctive nature as an atrocity. During the decades immediately after the war, there were some less publicized historical assessments. Historians today, Confino holds, face the neglected task of evaluating their historiographical importance.[7]

I begin with a contrast of historians Raul Hilberg and Saul Friedländer, two scholars of the same generation who set the tone of rising discussion from the 1960s. Both were Central European refugees from Nazi persecution. Both reflected on the monumental significance of the Holocaust in modern history. Both survived the precarious times of their youth to become leading historians of the Holocaust. Each was preoccupied with memory, but in different ways and for different reasons. Hilberg provided the first well-documented historical presentation of the reality of the genocide of European Jews. Friedländer provided the keynote to the issue of memory as an essential element of any historical interpretation of the nature and meaning of the Holocaust.

Raul Hilberg: A Historical Prelude to Historical Remembrance

Raul Hilberg (1926–2007) was an impressive man, and not as dour as critics have made him out to be. In his 37 years as a professor at the University of Vermont, he was respected as a conscientious teacher and a generous colleague. When I first picked up *The Politics of Memory* (1996), his autobiographical memoir of his journey as a historian of the Holocaust,

---

[7] Alon Confino, *Germany as a Culture of Remembrance: Promises and Limits of Writing History* (Chapel Hill: University of North Carolina Press, 2006), 21.

I was puzzled by the title he had given it.[8] The topic of memory seemed at odds with the hard-edged empiricism of his carefully researched history of the Holocaust, regarded by scholars as foundational in light of the evidence of Nazi war crimes that he had amassed. But as I read his memoir, I recognized that his turn to the issue of memory in his later years was a response to what he had been unable to give his readers in his imposing history when first published—an edifying interpretation of Jewish resistance to the Nazi plan for their extermination. His thesis as developed in his detailed *Destruction of European Jews* (1961) was pervasively grim, as all parties—perpetrators, victims, and bystanders—were caught up in the process of genocide when its mechanisms were considered in their ensemble.[9] His portrayal of the dark side of humanity—compliance and passivity in the face of a project sublime in its evil—was unrelenting. In Hilberg's critique, no one was spared. The Germans as a nation were responsible as perpetrators, for it was their civil service at every echelon that made possible the momentum of the killing machine carried out with few compunctions. In Hilberg's view, any attempt at a convenient divide between the compliant and the zealous in carrying out the genocide is contrived. Jewish victims, moreover, bore responsibility for having passively bent to the German plan for their own persecution and eventual extermination. Jewish councils temporized with the policies of their oppressors until it was far too late, ever hopeful that acquiescence to discrimination and fine-tuned distinctions concerning shades of German versus Jewish identity would spare the worst for at least some of those marked for persecution.

Hilberg was well-qualified to advance such a comprehensive interpretation. He himself had come to the USA as an adolescent with his family in flight from the Nazi takeover of their native Austria. As a soldier in the American army toward the end of the war, he was trained in intelligence and caught his first glimpse of the Nazi's campaign against the Jews.[10] Later, while a graduate student at Columbia University, he gained access to a larger record as an interpreter of documents for army intelligence.[11] For his doctoral thesis, he labored for years in the vast collection of German state papers confiscated by the American army and housed in the

[8] Raul Hilberg, *The Politics of Memory; the Journey of a Holocaust Historian* (Chicago: Ivan Dee, 1996).

[9] Raul Hilberg, *The Destruction of European Jews*, 3d ed. (1961; New Haven, CT: Yale University Press, 2003), 3 vols.

[10] Hilberg, *Politics of Memory*, 56.

[11] Ibid., 59–66.

US Archives in Washington DC.[12] In terms of the evidence he marshaled for his study, Hilberg's factual knowledge was thorough.[13] At the same time, he was sensitive to the rhetorical aspect of historical composition. His approach was shaped by his mentors at Columbia University, Solo Baron and Franz Neumann, who emphasized the workings of the bureaucratic apparatus of the German state in carrying out the vision of Hitler and his most doctrinaire henchmen. The Final Solution, Hilberg argued, was not a calculated plan from the beginning but emerged out of discriminatory practices put in place incrementally through the echelons of the German state apparatus. Along the way, respect for law eroded, giving way to arbitrary decrees and eventually to the chaos of direct verbal orders to party activists and fanatical vigilantes on the local level. Opportunism, more than ideology, made the machinery of destruction possible. The plan for genocide seemed possible because of the indifference of ordinary Germans, the compliance of Jewish leaders, and the willingness of too many ordinary Jews to succumb to their oppression in light of the passivity that had been the strategy of survival for their ancestors since the Middle Ages.

Hilberg won a prize for his dissertation from Columbia. But he faced difficulties in finding a publisher, and, once his work was published, stiff criticism from reviewers, notably those in the American Jewish community. His historical account, they argued, left no room for remembering the genocide in terms of heroic resistance on the part of victims. His was an account that put process over personality. The human face of those who suffered was absent from his narrative, his critics argued. Such criticism, coupled with professional jealousy, launched what Hilberg characterized as his "Thirty Years War" with his critics.[14]

As a response to critics who argued that his account dealt in impersonal processes and neglected individual personalities, Hilberg several decades later followed with *Perpetrators, Victims, Bystanders* (1992), a more concise and readable account of his historical interpretation.[15] Herein he paid more attention to the players in this historical episode. But his account was still framed around social categories of participation—

[12] Ibid., 71–73.

[13] See Hilberg's historiographical study on methods of evaluating evidence, *Sources of Holocaust Research; An Analysis* (Chicago: Ivan Dee, 2001).

[14] Hilberg, *Politics of Memory*, 123–37.

[15] Raul Hilberg, *Perpetrators, Victims, Bystanders; the Jewish Catastrophe, 1933–1945* (New York: HarperCollins, 1992).

the civil service, the liberal professions, the leaders of Jewish councils, and many less prominent agencies not only in Germany but throughout Eastern Europe. He illustrated his thesis with case studies of the behavior of participants, perpetrators and victims alike. This follow-up study only reinforced his thesis that this was a history without heroes or prospects of redemption. As he remarked in his memoir, this was a difficult message to convey because it did not accord with the way in which posterity wanted to remember this experience. This second book, therefore, was dismissed as derivative, and received almost no scholarly attention. Hilberg was respected, but rarely cited.[16]

Surprisingly, Hilberg had few rivals during the early years in which he carried out his research and published his findings, a task that required patience and perseverance in the face of professional and bureaucratic obstacles. Though he won praise and admiration from his doctoral professors and his academic peers, he could not give his critics a memory of those times that would satisfy a longing for some glimmer of redemption that they could take away from his interpretation. What was wanted was an edifying memory. That Hilberg would not give them. His account of the relentless process of destruction that compromised victims as well as perpetrators was judged too controversial in terms of the sensibilities it might offend. Accordingly, it was marginalized. Hilberg was acknowledged as the first major Holocaust historian. But Holocaust scholarship moved on, expanding its search for ways to find some positive meaning posterity could use to make sense of an atrocity of such magnitude. In the historiography of the Holocaust, the politics of remembrance would loom over historical narratives for decades in a quest for an interpretation with some consoling meaning that posterity could use.

SAUL FRIEDLÄNDER ON THE LIMITS OF REPRESENTATION

Memory in Holocaust studies is now a vast field of scholarship. But a good place to begin is the early work of the Israeli historian Saul Friedländer (b 1932). In his writings as in his life, he illustrates the psychoanalytic approach to memory studies that developed in the scholarship of the 1980s. For him, the unrequited memory of Holocaust survivors was the essential yet elusive element for any comprehensive interpretation of the Holocaust in German history. As a Holocaust survivor himself, the

[16] Ibid., 161–75.

personal probing of his own memories of the war years had been a resource for unraveling his own complex identity, and would play into his method as Holocaust historian. *When Memory Comes* (1979), his psychoanalytic memoir, recounts his survival as an adolescent in Vichy France during the war years under an assumed identity, and his immigration to Palestine after the war to take on a new identity in the incipient Israeli nation. His memoir portrays the way his memories of childhood came back in middle age in fragments, little by little.[17] The need of survivors like Friedländer to work through the trauma of their ordeal gave new life to Freud's theory of psychoanalysis. The project of working through repressed memory as a preliminary step to writing histories of the Holocaust served as a guiding principle of scholarship in this field during the last decades of the twentieth century.[18]

As a child, he and his family had fled their native Czechoslovakia on the eve of the war and found refuge in Vichy France. It provided only a temporary sanctuary. His parents were arrested and sent to the death camps. But before their departure they succeeded in placing him with a French family under an assumed identity. He grew up Catholic, excelled in the French educational system, and might have made a good life for himself there. But having been persecuted for a Jewish identity of which he knew little as a child, he decided as he came of age after the war to embrace that heritage by starting life anew in the pioneering experiment of the fledgling republic of Israel. Years later, settled in his role as professor at Hebrew University in Jerusalem, Friedländer paralleled his memoir with a scholarly inquiry into methods for the study of historical psychology.[19] Together they prepared the way for the approach he would take in his contribution to the study of the Holocaust, for which he is known as a leading scholar. In searching his past, he was coming to terms with multiple identities, and so recognized the central role of memory in his own formation as historian. The relationship

[17] Saul Friedländer, *When Memory Comes* (New York: Farrar, Straus and Giroux,1979).

[18] A provocative counterpoint to the psychoanalytic approach to the memory of the Holocaust was offered by University of Chicago historian Peter Novick in his *The Holocaust in American Life* (Boston: Houghton Mifflin, 1999), esp. 1–15. He introduced issues about the politics of memory into the discussion. Employing the Halbwachian model, he argued that Jewish-American leaders, responding to the identity politics of the 1980s, publicized Holocaust commemoration strenuously for fear it might otherwise be crowded from public attention.

[19] One of Friedländer's first books addressed this Freudian-inspired historiography, *History and Psycho-analysis* (New York: Holmes & Meier, 1978).

between memory and historiography in scholarship on the Holocaust came to assume a larger role than the history itself.

Over the course of the 1980s, Friedländer canvassed the growing body of scholarship on the history of the Nazi years in Germany, with an eye to its historiographical implications.[20] He played a major role in the Historians' Dispute. By then, a considerable body of scholarship on the subject had been amassed, and German scholars were calling for its integration into a comprehensive narrative of German history. But in his review of this scholarship, Friedländer noted a paradox. The further the project of historicization proceeded, the more problematic became the prospect of fashioning a master narrative of the era. Historians found themselves at an impasse. Many sound interpretations were offered, though their perspectives varied widely, and no coherent overview seemed in the offing. For Friedländer, the linchpin of Holocaust interpretation depended upon reckoning with the historical meaning of the Final Solution, the Nazi's calculated plan for the extermination of European Jews. Its reality defied conventional historical explanation.[21]

German historians agreed about the hateful crimes of the Nazis and that historians had an obligation to assess their nature and meaning within the context of the era. But the limits of responsibility beyond the perpetrators remained unresolved, and many historians sought to interpret the experience within the context of traditional historiographical categories, often by placing the 12 years of the Third Reich within a longer historical timeline. Some historians stressed the deep ideological origins of National Socialism; others turned to structuralist interpretations that stressed the impersonal workings of state bureaucracy, whose authoritarian cast had been fashioned in the late nineteenth century by Chancellor Otto von Bismarck. Among conservative scholars, the Holocaust was one wartime atrocity among many. The crimes committed by the Nazis, while heinous, were, nonetheless, comparable to others, notably those committed by Stalin. They pointed to the plight of Germans living in Eastern Europe in the face of the advancing Red Army. There were issues of wartime diplomatic and military strategy to consider as well. For scholars with left-wing leanings, by contrast, capitalism was the villain, spawning fascism as its

---

[20] Saul Friedländer, *Memory, History, and the Extermination of the Jews of Europe* (Bloomington: Indiana University Press, 1993), is a collection of his scholarly articles on the topic over the previous decade.

[21] Ibid., 22–23, 55, 102–04.

most perverse manifestation. The intent of all such discourse, Friedländer contends, was to "normalize" the experience of the Nazi era by evaluating it within conventional frameworks of historical interpretation.[22]

Friedländer found particularly interesting the argument advanced by historian Martin Brozat, who issued a plea for the historicization of National Socialism on cultural rather than political grounds. Brozat challenged the tendency toward a monumental interpretation of the era, as if it were a drama between forces of good and evil. The era was too complex for such moralizing, he contended. He condemned Nazi war crimes but proposed an interpretation that narrowed the circle of responsibility to those actively engaged in committing them. For the majority of ordinary Germans, he argued, everyday life was not conspicuously different from what it had been in earlier years. There were as well modernizing trends in German culture and industry that could be appreciated within a continuum of history that ran through the Nazi era. Politics and ideology could not explain all that transpired during that time period.[23]

In Friedländer's view, Brozat, and for that matter most German historians of the Nazi era, evaded a full confrontation with the reality of the Final Solution. Most of the scholarship had glossed over psychological issues. But Nazi rhetoric had been apocalyptic and mythical, Friedländer argues, and their deeds strayed beyond the bounds of rational explanation. Historical interpretation of the Holocaust, therefore, had arrived at an impasse and historians were confronted with memories of victims and perpetrators that could not be explained away. The only recourse was to take traumatic memory seriously. The meaning of the Final Solution remained embedded in the memories of those caught up in its workings, hidden for the most part in repression, whether it be the guilt of perpetrators or the anguish of victims. Memory was the key to any master narrative of the Nazi era, and yet such memory was opaque and inaccessible. The past as memory, therefore, had not faded as memories typically do, but rather festered openly as the presence of an unrequited past. Anyone who seriously studied the era, Friedländer allows, was aware of uncanny memory, promptings for addressing the unresolved issues repressed for more than 40 years. Repressed memory, therefore, blocked the possibility of historicizing the era in a master narrative. For this reason, one had no choice but to treat the Holocaust as exceptional and unprecedented. Ordinarily, his-

---

[22] Ibid., 51, 69.
[23] Ibid., 35–36, 65–67, 75–79.

tory took charge of memory through critical analysis. But in this instance the memory of the genocide of European Jews stood out as a persistent presence in its impenetrable autonomy.[24]

For Friedländer, moreover, there was an urgency about addressing the memory issue. Holocaust survivors, the existential bearers of that memory, were aging. Within a few years, all would have passed away, leaving ritual atonement and historical interpretation as insufficient instruments for exposing the full meaning of the Holocaust. Regrettably, he concludes, there may always be limits to the historical representation of this episode in history, a gap in the historical record marked by a place of unrequited memory.[25]

With interpretation by professional historians immobilized by unresolved issues of memory, Friedländer concedes that the "culture industry" had assumed the primary role in representing the Holocaust for the public at large. Films, television series, memoirs, and popular novels, deeply appreciated by German and American audiences, were assuming the preponderant role in shaping Holocaust remembrance. In all of these interpretations, he notes, the present had come to displace the past as the referent for understanding the Holocaust in history. Decades after the event, moreover, renewed discussion about the meaning of the Holocaust was introduced into an age markedly different from the one in which the Nazis had committed their crimes. How many today, he wondered, continued to care deeply about those times. Many of the older generation of Germans were indifferent, while the younger one had moved on to the preoccupations of their own times. The context for consideration of the history of the Holocaust was no longer the nation but the world at large, a setting with new realities and its own set of problems. Still, for Friedländer, historians had an obligation to set the record straight, to persevere in their efforts to get at the truth about the reality of that past.[26]

Friedländer's historiographical meditations, therefore, run against the grain of postmodern historiography, which is relativist and present-minded. Many of the realities, hidden in unreported memories of victims, continued to await exposure. Historical scholarship could not close the book on their experience until all of its possibilities had been ferreted out and discussed publicly. It was with this understanding that he undertook his

[24] Ibid., 2–4, 16–17, 48, 103–20.
[25] Ibid., 47–57, 120–30.
[26] Ibid., 7, 47, 59, 61, 85–100.

own comprehensive history of the Holocaust, hoping to give voice to victims as well as an account of their suffering.[27] Still, he admitted, it is unlikely that there will ever be complete closure in the certainty of scholarly consensus.[28]

EVA HOFFMAN: MEMORY AND THE CHILDREN
OF SURVIVORS

Friedländer's psychoanalytic approach to the traumatic experience of Jewish victims of the Holocaust correlates with that of essayist and literary critic Eva Hoffman (b 1945). A Jewish child born in the last days of the war, she wrote a memoir about the long-term effects of the Holocaust on her parents, who survived in hiding and after the war migrated to Canada.[29] She explains the staying power of the memory of the Holocaust, as it continues to trouble the conscience of the present age. There are many memoirs by Holocaust survivors. Hoffman's is distinctive for its perspective on the role of their children as bearers of their memory as it is absorbed into history. A psychologist by training, she recounts her perception of what the experience did to them, and her assessment of its influence on her own life, and as a crime against humanity, on all of us. She contributes to our understanding of the way the Holocaust over time passed from personal to collective memory, taking on a historical identity in the process. Her memoir might be thought of as a way of honoring her parents by putting their story in a larger perspective some 60 years after the event. She wanted to fulfill her sense of filial responsibility to them.

For the children of survivors, as for the survivors themselves, coming to terms with the meaning of the Holocaust required time. Her parents

---

[27] Saul Friedländer, *The Years of Extermination: Nazi Germany and the Jews, 1939–1945* (New York: HarperCollins, 2007), 870 pp.

[28] Friedländer, *Memory, History, and the Extermination of the Jews of Europe*, 62, 123, 133–34. For the continuation of the memory issue emerging out of Holocaust studies, see Wulf Kansteiner, *In Pursuit of German Memory; History, Television, and Politics after Auschwitz* (Athens: Ohio University Press, 2006); Andreas Huyssen, *Present Pasts; Urban Palimpsests and the Politics of Memory* (Stanford: Stanford University Press, 2003). On further reflections on the Vichy Syndrome, see Richard J. Goslan, ed., *Fascism's Return; Scandal, Revision, and Ideology since 1980* (Lincoln: University of Nebraska Press, 1998), 182–99. For the Holocaust in Poland, see Jonathan Huener, *Auschwitz, Poland, and the Politics of Commemoration, 1945–1979* (Athens: Ohio University Press, 2003).

[29] Eva Hoffman, *After Such Knowledge; Memory, History, and the Legacy of the Holocaust* (New York: Public Affairs, 2004).

never told their story to anyone outside their family. She takes on that task by relating their experience to her own psychological reckoning with it: of understanding their story in relationship to her own, of forgiving herself for setting aside her family's troubled past as she pursued her own ambitions in young adulthood. Her account is informed by psychoanalytic insight, but in the end she turns to historical analysis. If Hoffman shows us how the legacy of the Holocaust persists into our own present age, and in the process how its collective memory has evolved. Looking at the structure of her memoir, one sees her interpretation of the life cycle of the memory of the Holocaust, as it passed from a family fable toward historical understanding. She evokes memories of her childhood and of her parents that only she can understand subjectively. But she also shows how her family's experience has come to play into our collective memory of the Holocaust, and she ruminates on the implications.

Hoffman was born in into an orthodox Jewish family in Cracow, Poland, in 1945, shortly after the end of World War II. Before and during the war years, her parents had lived in a small town in what was then a Ukrainian-speaking province of Poland. They spent the war years hiding in an attic, and so were among the fortunate few who escaped the Nazi dragnet. Several of their relatives were captured and sent to the death camps. When Hoffman was 14 years old, her family immigrated to Vancouver, British Columbia. Her parents brought with them the memories and the mores of the *shtetl* culture of Eastern European Jews. But Eva made her own way by assimilating readily to her new secular, middle-class, suburban surroundings. As for immigrants in preceding generations, Canada, like the USA, was a land of opportunity, and Hoffman used her talents to succeed in her new country. A gifted student, she did graduate work, taught, and became a writer.[30]

Hoffman's memoir serves as a late-life reflection on her path into adulthood in light of her heritage as a child of Holocaust survivors. She wrote it in order to reexamine her parent's lives before she was born, her own childhood, and her coming of age. In the process, she came to see how deeply the Holocaust had shaped her life in unconscious ways. Hoffman understood her role vis-à-vis the Holocaust to be that of a mediator between her parents, her own generation, and posterity. As a child, she internalized the suffering of her parents. Their past was for her an unhappy tale of vague impressions with mythical overtones. As a mature adult, she sought to

[30] Ibid., 82–84.

explore the psychology of her relationship to her parents in all that they had endured. Her parents died in 1991, and with them their personal memories of their ordeal. Hoffman, therefore, sensed her responsibility to contribute their story to the recorded testimony of the Holocaust.[31]

Two themes sustain Hoffman's narrative. The first concerns her physical journey. Her story begins with her emigration as an adolescent to a new land and a new kind of life. It ends with her pilgrimage late in life to the place of her parent's ordeal. The second addresses her psychological journey, from mythical to historical understanding. For survivors, the Holocaust was inextricably tied to its aftermath. Though liberated from the death camps, many of them had no home to which to return. Their stories after the liberation were about flight, their need to escape the place of their suffering to some better refuge. From a historical perspective, their migrations may be thought of as a modern Jewish diaspora. Survivors found refuge around the world. Many made a new life in the USA or Canada. A smaller number made their way to the new Republic of Israel. Starting over in an unfamiliar land was never easy. Most émigrés had few material resources and limited moral support in their new surroundings. In all cases, emigration meant further dislocation. The trauma of return to the world followed upon the trauma of having been wrested from it.

Nor was America a particularly hospitable safe haven for Holocaust refugees. It is not that Americans were without compassion for these victims of atrocity. But in postwar America, most people were getting on with their lives and ready to put the wartime past behind them. America was in the midst of an economic boom, and Americans were bent on pursuing their materialist dreams of owning suburban homes, kitchen appliances, new cars, and leisure pastimes. The secular culture of individual initiative in America was a world apart from the closed, interdependent *shtetl* culture of Eastern European Jews. Hoffman felt the conflict between her need to take advantage of her opportunities and her responsibility to shelter her parents, who never fully adapted to their new homeland. As an adolescent, she honored her parents but chose the modern secular American culture. It was possible for an immigrant child to assimilate readily and to thrive. American culture favored the best of liberal values in its stress on self-reliance as the route to success. It was optimistic, energetic, resourceful, and hopeful about the future. By contrast, the *shtetl* world of her parents

---

[31] Ibid., 27–28, 90–97, 92, 96–98.

had been cautious, deferential toward parental authority, suspicious of outsiders, clanish in its strategy for survival.[32]

As an adult, Hoffman made trips to the lands she associated with the Holocaust past of her parents—Germany, Israel, and the small town in the Ukraine from which her parents had come. The first two destinations were for academic conferences. The last, to Zalośec in the mid-1990s, was a pilgrimage to the place of her parent's earlier lives in search of knowledge of its historical realities. This town was no longer in Poland, but in the Ukraine. Most of the Poles had fled; all signs were in Ukrainian. But a few of the residents who had once known her parents still resided there. Some remembered them, and recounted specific facts about their lives there. For Hoffman, this was the final act in demythologizing the fable of her heritage. Historical knowledge helped her to put her own past in a broader context.[33]

In exploring her psychological journey, Hoffman explains the long process by which she became aware of the relationship between her knowledge of her parents' suffering and the larger event which since the 1970s we have come to call the Holocaust. Here we see her psychoanalytic bent as she works through her unconscious memories en route to raising conscious moral and historical issues. It was some time before she became aware of the psychological burdens she carried.[34] The stages of her journey follow her chapter headings: from myth, to psyche, to ethics, to memory, to history as research, to history as interpretation.

To come to terms with her parent's past meant confronting the moral issues the Holocaust raised. As she came of age, Hoffman awakened to their gravity for posterity. First and foremost were questions of how and why the Holocaust had taken place. How could one explain the reality of evil in a supposedly benevolent civilization? For Hoffman, there is no explaining it away. The German Nazi elite had committed crimes against humanity, and they had done so without remorse. The best one could do was to try to understand why. For Hoffman, the Nazi perpetrators were able to carry out the mass extermination of their victims because they dehumanized them. They treated Jews as it they were subhuman. As the alien Other, Jews could be liquidated without qualms.

---

[32] Ibid., 80–89.
[33] Ibid., 127, 151, 205–20.
[34] Ibid., 55, 60, 70–74.

For Hoffman, the Nazi perpetrators bear primary responsibility for the Holocaust. But they did not act alone. Their plan was possible because of the passive complicity of the majority of the German people. As a consequence, it was difficult for her to overcome her perception of Germans as the Other. In the course of her work as an adult, she came to know many Germans of her own generation, and she did not hold them personally responsible. But at a deep psychological level, her sense of distance from them persisted. She recognized, of course, that the children of the perpetrators were victims, too, and had to work through their understanding of their parent's active or passive complicity in the Nazi crimes.[35]

From ethical issues Hoffman moves to those of the collective memory of the Holocaust. Strategies for reconstructing that memory entered public discussion during the 1970s, when the media began to publicize the event in movies and politicians in official rituals of commemoration. It is in the invention of this collective memory that the concept of the Holocaust took shape. Hoffman is suspicious of the idea of collective memory, conceived as a shared consciousness. What we call "memory," she suggests, is not memory at all. Authentic memory is personal and subjective. Collective memory is not a shared subjective consciousness of past experience, but a set of attitudes articulated by journalists and politicians. Such a memory is at best a second-hand reconstruction of personal memories that serve present political needs. Collective memory distorts the past because it is reductionist, stereotypical, and easily manipulated. It conforms to the political requirements of the day. It invokes the past to serve other purposes.[36]

It was at this juncture that Hoffman felt capable of writing about her understanding of the place of the Holocaust in her family's history. She saw historical research as a corrective—a way to test her memories against specific realities she was able to uncover about her parents' past. Fashioning a collective memory, she suggests, is always a negotiation among present needs. Historical research, by contrast, offers the advantage of trustworthy evidence. It was in search for such evidence that she made her pilgrimage to the Ukrainian village from which her parents hailed. As she put it,

---

[35] So, too, Hoffman felt a distance from Catholic Poles. They had suffered too in their defeat and occupation by the German army. Though some three million Polish Jews died in their campaign, so too did three million Polish Catholics. After the war, they did little to acknowledge the suffering of their Jewish brethren. They showed no enthusiasm for honoring their memory in public acts of commemoration. Ibid, 14, 106–11, 117–33, 136–4.

[36] Ibid., 66, 156–58, 160–70, 174–77, 180–81, 185–90, 229.

she needed to find a historical context for interpreting their experiences adequately.[37]

Finally, Hoffman turns to the issue of the larger historical context in which the Holocaust may be situated. For some students, the Holocaust, a crime against humanity, was a unique event. In its planning and execution, most would agree that as a calculated act of mass atrocity carried out with industrial efficiency it was unprecedented. But Hoffman sees the interpretative value of comparing it with the trauma that others have suffered in other twentieth-century crimes of genocide—the Armenians in Turkey, the Tutsis in Rwanda, the Muslims in Bosnia. Hoffman's point is that the suffering of victims is subjective and personal. But genocide itself is an attack on our shared humanity. We need to understand it and to learn from it.[38]

GABRIELLE SPIEGEL AND PAUL RICOEUR: HISTORY VIS-À-VIS HOLOCAUST MEMORY

Gabrielle Spiegel

As a fitting postscript to all the work on memory and history in Holocaust studies, historian Gabrielle Spiegel has offered an insightful perspective on the tendency among some scholars to accord to memory the status of history.[39] Drawing on the work on Jewish religious tradition by Yosef Yerushalmi, she sets memory and history apart on the basis of their opposing conceptions of historical time. Collective memory evokes the presence of the past. Particularly in its ritual expressions, it contributes to a sense of reliving the past as an act of renewal. Its understanding of time is accordingly cyclical. History, by contrast, establishes a distance between past and present. It insists on the singularity of events that occur but once and for all time. As a conception of time, therefore, history is linear.

Spiegel pursues this distinction in two contexts of Jewish religious thought: that of the Middle Ages and that of the era of the Holocaust

---

[37] Ibid., 180, 192, 196–98, 242–43, 270–75. See also Marianne Hirsch, *The Generation of Postmemory; Writing and Visual Culture after the Holocaust* (New York: Columbia University Press, 2012), who pursues the theme of legacies of remembrance from a broader historical perspective.

[38] Hoffman, *After such Knowledge, 112,* 158–60, 178–79.

[39] Gabrielle Spiegel, "Memory and History: Liturgical Time and Historical Time," *History and Theory* 41 (2002): 149–62.

and its aftermath. In the earlier era, the distinction between memory and history was clear. Religious Jews thought in terms of sacred time, a realm apart from historical time. All events, even those of suffering and unhappiness, were integrated into the cycles of Jewish tradition in hope of their eventual redemption. In the era of the Holocaust, such a conception of the redemptive quality of collective memory within tradition ceased to offer consolation to its survivors. The Nazi project of genocide included the obliteration of all memory of the Jews.[40] It was a historical event impossible to integrate into the sacred time of remembered tradition. Holocaust memories, therefore, took on the singularity of history, "fugitive memories" outside living tradition and history alike. The strategy of some Holocaust historians, Spiegel contends, was to privilege the privatizing of such memories. For Holocaust survivors, their memories stood apart from the collective memory of shared faith. In the memory/history controversy, therefore, the possibility of historical understanding was compressed between inaccessible private memories and historiographical discussion of the limits of historical representation.

Spiegel further notes that this discussion of the Holocaust between memory and history permits comparison with Nora's project on the French national memory. Nora, too, addressed collective memory at the end of and apart from the traditions in which it was once immersed. Memories divorced from living tradition cannot bind people collectively but rather isolate them individually. They come to be perceived as discrete episodes, belonging neither to tradition nor to history, but only to topical places in historiographical schemes of the sort that Nora had devised. The question is whether the structures of time that appear as constants in Spiegel's interpretation have since been recast in the historiography of the memory phenomenon. Memory studies point to the destabilization of the structures of historical time, drawing history closer to memory in the presentist perspective toward which it gravitated in the late twentieth century. This perspective has led to reflection on the mnemonics of historical time.

Paul Ricoeur

The French phenomenologist Paul Ricoeur complements Spiegel's historiographical interpretation by adding a philosophical dimension. He took

---

[40] See Pierre Vidal-Naquet, *Assassins of Memory; Essays on the Denial of the Holocaust* (New York: Columbia University Press, 1992), 57, 102.

up the topic of memory for his last major book, *La Mémoire, l'histoire, l'oubli* (2000), inspired by the intense preoccupation with memory as an object for critical inquiry in the late twentieth century.[41] The memory of the Holocaust haunts his narrative, for it remains too close to our worries about man's capacity for inhumanity toward his fellow man to be set aside. He strives to remain faithful to its memory by keeping the realities of history in mind. As he explores the many routes into the history of memory, he returns continually to his thoughts on the question that so preoccupied a generation of historians of the Holocaust: how may we reconcile remembrance of a traumatic event (acknowledging the past that was) with the need to move on (caring about a past that is no longer). An event as infamous as the Holocaust, he cautions, may find consolation in memory only by acknowledging history's truth.

Ricoeur turns to the history of the Holocaust as a context for understanding memory's vocation—its faithfulness to the existential reality of the past. Bearing the imprint of grief and distress, memory of the Holocaust provides a bulwark against the assaults of those who would deny its reality. Herein Ricoeur suggests the significance of efforts to record the testimony of Holocaust survivors, a task that took on such urgency in the late twentieth century. The project insured that remembrance of the crimes of the Holocaust involved more than could be contained in the historical representation of its atrocities. In its fidelity to the living memory of survivors, memory serves as a guarantor of its realities, an anchor in the past that enables historians to get at the truth about that experience in the variety of its resources for its representation. But history must respect the mourning that memory of the Holocaust engenders. It must first beg forgiveness (*pardon*) of the memory of the Holocaust before proceeding to an analysis of its meaning, an activity that distances interpretation from living witness.[42]

Ricoeur follows with a question about what was a central philosophical issue in Holocaust remembrance: is it possible to forget an unhappy memory? Here he explores a counterpoint to the limits of representation—the limits to the possibility of lifting the veil of its repression. Politicians, he points out, have adopted many strategies to redirect attention from unhappy memories. In antiquity, he notes, the concept of "greatness" was understood as the political capacity to impose peace where discord

---

[41] Paul Ricoeur, *La Mémoire, l'histoire, l'oubli* (Paris: Seuil, 2000).
[42] Ibid., 593–95, 646–50.

reigned. Discord may be placated, but can an unhappy memory be overcome? Does amnesty, he asks, not wear the mask of amnesia? Here Ricoeur alludes to the "subterranean empire" where strategies for forgetting hold court. He reviews their arts: symbolic compromise, reparations, repentance, and absolution. Forgetting in one of these modes becomes a political act, domesticating the meaning of the Holocaust while honoring its victims in commemorative rituals. But the discord such atonement is meant to assuage lies deeper, Ricoeur contends, for the memory on which it is based can never be completely requited. Memory as guardian of the past remains ever vigilant. A memory that is forgiven does not mean that it is forgotten. Therefore, he concludes, we cannot speak of happy forgetting in the way we speak of happy remembering. As Friedrich Nietzsche pointed out, there are instances in which we exercise our will not to forget in our effort to remain faithful to the past. In this respect, Ricoeur concludes, there is an asymmetry between remembering and forgetting. Remembering is a particular experience, a moment of illumination. Forgetting, by contrast, is a process that plays out over time and that resists completion. Some experiences are impossible to forget in light of our reserves of caring. To believe that one can forget, he concludes, is a "barbarous dream."[43]

Jeffrey Olick on Symbolic Compromise in the German Politics of Holocaust Memory

Sociologist Jeffrey Olick (b. 1964) explores the way in which Ricoeur's formulae for forgetting have played out in the practice of German politics. He has traced the way the West German government over the course of nearly a half century struggled to integrate acknowledgment of the evils of Nazi atrocities into formal rituals of atonement, a symbolic reckoning with Germany's past calculated to spare younger generations from association with the crimes of their ancestors. He characterizes their project as a "politics of regret." [44]

To illustrate his argument, Olick concentrates on mnemonic practices in the strategies employed by statesmen of the Federal Republic of Germany over the course of its first 45 years, for no case is more poignant in revealing

---

[43] Ibid., 650–54.

[44] Jeffrey K. Olick, *The Politics of Regret; On Collective Memory and Historical Responsibility* (New York: Routledge, 2007).

the difficulties of coming to terms with the memory of the Holocaust. He takes 8 May 1945 as his point of departure, the day the Allies won Nazi Germany's unconditional surrender. From the outset, he explains, there was an underlying ambiguity about its meaning in postwar commemorations. Was 8 May a day of defeat or one of liberation? To choose the former might be construed as an affront to German patriotism. To opt for the latter might be viewed as an attempt to evade responsibility for this nefarious past. Framed as an irreconcilable opposition between ways to remember the last day of the war in Europe, the lines of tension its commemoration generated remained taut for decades. As Olick puts it, public discourse on the subject was taboo for the generation that had come of age during the war years. Only for the generation of the 1960s was the moral reckoning with the Holocaust opened for widespread public discussion with the trial of Adolf Eichmann. During the 1980s, the Historians' Dispute prompted statesmen to devise a new strategy for reconciling present-day Germany with the tainted legacy of its past.[45] To trace the history of the official commemoration of 8 May as the end of a regrettable era, Olick explains, illustrates the nature of the standoff between generations en route to a symbolic compromise.[46] As such, the politics of regret was to become a new "principle of political legitimation."[47] Though he does not employ psychoanalytic language, Olick's argument about symbolic compromise is reminiscent of that offered about screen memories by Sigmund Freud in *Totem and Taboo* (1913). Intolerable memories of violence are repressed and find conscious expression only in more tolerably benign images. The existential horror of the private memories of individual victims of the Holocaust, therefore, would be brought to the surface of public memory only in commemorative disguise. What had once been a repressed taboo about discussing the misdeeds of the wartime generation emerged into consciousness as a ritualized prohibition, to serve as an admonition for future generations.[48]

---

[45] Ibid., 55–83, 139–51.

[46] To make his case, Olick formulates a theory of genre memory. For him, a genre is the commemorative matrix of a memorable event. The first commemoration sets the chronological track that subsequent ones will follow. The memory of that event is modified over time, but there is a logic to the way it evolves into more abstract and idealized images of remembrance. In his terminology, the trajectory of the periodic commemoration of an event is "path dependent" upon its first formulation. Ibid, 55–57.

[47] Ibid., 14, 121–38.

[48] Ibid., 50–53.

In tracing the path of the commemoration of 8 May 1945, Olick identifies three successive modes of official remembrance.[49] During the 1950s, German Chancellor Konrad Adenauer endeavored to create an image of Germany as a "reliable nation," its people saddened by the way its governance had been hijacked by Hitler and his henchmen, and now ready and willing to comply with the demands of the victorious Western allies that they rededicate themselves to the democratic ideal. In fashioning an official memory of the meaning of 8 May, Adenauer put his accent on the culpability of the few in order to spare the many ordinary Germans from guilt by association. During the following decade of the 1960s, Olick contends, the government moved left under the leadership of the idealist Willy Brandt, who chose the high ground of portraying Germany as a "moral nation" willing to acknowledge its sins and its incumbent responsibilities in light of them. Olick recalls Brandt's now famous personal gesture of atonement in the name of his nation, as he spontaneously knelt at the monument commemorating the uprising of the Warsaw ghetto in which so many Jews sacrificed their lives in a desperate act of rebellion. Here emerged the first stirrings of the official politics of atonement, to be practiced through the ritual remembrance of the horrors of the genocide of European Jews at the hands of a German regime. But there was a politics to this remembrance, too. For Brandt, accepting responsibility was also a means of asserting a measure of German independence vis-à-vis the American-dominated Western alliance, and of seeking overtures with the Eastern Bloc in the interest of promoting a lasting peace in Europe. The periodic commemoration of 8 May during the 1970s and especially the 1980s, in turn, signaled the desire of more conservative German statesman such as Walter Scheel and especially Helmut Kohl in his long tenure as chancellor to lift the burden of atoning for the memory of the Holocaust by placing its commemoration in a broader historical context. Kohl's political purpose was practical. He wanted to institutionalize the commemoration of 8 May so as to permit a younger generation to move forward with the project of rebuilding the pride of the nation as it entered the twenty-first century. The Kohl government made official apologies to Holocaust victims and would continue to do so in annual ceremonial observances. But these symbolic acts of atonement tacitly exonerated

---

[49] 109–15.

ordinary Germans, nearly all of whom by the 1990s were too young to have been guilty bystanders anyway.[50]

DANIEL LEVY AND NATAN SZNAIDER ON THE GLOBALIZATION OF HOLOCAUST MEMORY

I close with a discussion of the book by Daniel Levy and Natan Sznaider, *The Holocaust and Memory in the Global Age* (2006), which provides an overview of the way Holocaust remembrance has played out in a variety of trajectories in times closer to our own.[51] They argue that the reconfiguration of the collective memory of the Holocaust over time illustrates the larger process of transition from a European into a globalizing culture during the late twentieth century. As such, it lies at the heart of the larger interest in memory during that era. The facts of the Holocaust had long since been settled and were beyond dispute. But the question of how it ought to be remembered remained open through the debates of the late twentieth century and into our own times. The general pattern of remembrance, they argue, follows a move from memories that were particular in an experience judged exceptional toward those that are abstract and universal.

Levy and Sznaider employ two strategies for addressing the topic. They retrace the phases of remembrance from troubled repression on the part of victims and perpetrators in the aftermath of World War II to edifying lesson for posterity everywhere by the end of the twentieth century. But they also reconstruct the changing meaning of the genocide as it moved out of living memory into historical symbolization. They trace this reconfiguration of Holocaust remembrance by placing the topic within a framework that distinguishes its reception within contrasting historical epochs—what they characterize as the first and second modernity. (The distinction corresponds roughly to the more commonly employed modern/postmodern divide.) The transition from first to second modernity was an ongoing process during the second half of the twentieth century, though 1989, the date marking the end of the Cold War, serves as its symbolic threshold.[52]

---

[50] Ibid., 68–80, 100–01, 110–13, 142–43.
[51] Daniel Levy and Natan Sznaider, *The Holocaust and Memory in the Global Age* (Philadelphia: Temple University Press, 2006).
[52] Ibid., 6, 126–27, 179–88.

Collective memory in the first modernity focused on the nation-state as the primary referent of collective identity. In the second modernity such memory was reconfigured to encompass new conceptions of identity in a globalizing culture. Here Levy and Sznaider note a paradox. Memory in the second modernity operates dialectically in the interaction between the global and the local ("glocalization" is the term they coin) and in some measure rivals national memory. First modernity memory nurtured binding emotional ties between past and present and strove for unity in promoting pride and patriotism in allegiance to the nation-state. By the 1970s, the perception of the past such memory evoked was beginning to weaken. The Historians' Dispute of the mid-1980s was the German analogue for the crisis of national identity to which Pierre Nora had called attention for France. That crisis signaled the end of the first modernity. During the first modernity, the memory of the Holocaust was largely repressed and discussion of it enshrouded in taboo. Second modernity memory, by contrast, was self-reflexive. By this term, Levy and Sznaider mean that a naive subjectivity about memory gave way to a critical perspective on its workings. For Holocaust remembrance, reflexivity signaled the passing of the living memories of witnesses into historic symbolization of its meaning for posterity. In crossing that threshold, the Holocaust came to signify the plight of its victims as a referent for all who suffered in like circumstances in times closer to our own, even if the scope of persecution was not of the same magnitude. The atrocities committed by Serbian nationalists in Bosnia in the name of "ethnic cleansing" during the 1990s gave evidence of resurgent racism reminiscent of the Holocaust as a crime against humanity.[53]

First modernity Holocaust remembrance coalesced slowly, held in abeyance by the trauma suffered by its survivors. Their psychological distress made it difficult for them to recover and articulate their repressed memories of their ordeal. Holocaust memory, laden with anxiety and guilt, festered in the unconscious minds of its victims. Insofar as it was possible to gain access to these memories, recollection required long and painful psychoanalytic "working through" a confrontation with the realities of their experience. Respect for the plight of victims set limits on the interpretation of the historical meaning of the genocide. Discussion of the Holocaust was enshrouded in taboo and entered a long period of psychological latency. Meanwhile, European society moved on to the

---

[53] Ibid., 4–10, 26–30, 35–36, 39, 44–46, 127, 179–83, 191–203.

tasks of postwar reconstruction. Only gradually over decades following the war was the taboo partially lifted. Levy and Sznaider trace its tenuous beginnings to the publicity attending the capture of Nazi henchman Adolf Eichmann and his subsequent trial in Israel during the early 1960s.[54]

Open discussion of the meaning of the Holocaust was furthered by both German statesmen and historians during the 1980s. The Federal Republic of Germany instituted official acts of ritual commemoration and atonement, designed to acknowledge German responsibility for the Holocaust while sparing the younger generation of complicity in the sins of their forefathers. Significant, too, was the Historians' Dispute, in which professional historians wrestled with questions of representation and historicization of the genocide. In this respect, Levy and Sznaider contend, they saw their role as that of necessary intercessors between the victims and the public at large. Some argued persuasively that any attempt to integration of the Holocaust into historical narrative was premature. Too many survivors had yet to tell their stories, and there were limits to the representation of the trauma they had experienced. The historians set about collecting survivors' testimony, but cautioned that composition of a comprehensive narrative of the Holocaust should be postponed until they had considered fully how and to what degree the experience could be represented. The effect, Levy and Sznaider conclude, was to shut down interpretation of the larger historical meaning of the genocide. The Holocaust was judged exceptional and unprecedented, and so off-limits to historical comparisons.[55]

The reformulation of Holocaust remembrance in second modernity memory, Levy and Sznaider contend, moved discussion of its meaning from a German into a global context, opening the way for the public to understand the Holocaust as an emblem of extreme suffering in atrocities wherever they arose. The Holocaust exemplified a kind of suffering with which a public untouched directly by the Holocaust could empathize. Such privatization made manifest the diversification of Holocaust remembrance by the turn of the twenty-first century. It provided a route for lifting the taboos on its interpretation that historians had imposed in their conception of the limits of representation. Because it stressed emotional

---

[54] Levy and Sznaider note the writings of Hannah Arendt about Eichmann's psychology and that of confederates who joined him in pursuit of the extermination of European Jews. Ibid, 16, 42–43, 142, 57–95, 105–08.

[55] Ibid., 52–53, 68–81, 102–05, 120–21.

response to the suffering of others, second modernity memory permitted a more direct personal connection to Holocaust victims than the historians' guarded approach to interpretation had previously allowed. The realm of compassion for the suffering of strangers expanded, though it did not demand much in the way of active involvement in redressing old wrongs.[56]

Levy and Sznaider address the consequences of this reconfiguration of the historical remembrance during the second modernity. They cite several factors that contributed to remaking Holocaust memory as a symbolic frame of reference. The first was a changing understanding of the history of the Jewish diaspora. During the first modernity, their long history of exile and isolation rendered Jews aliens in a modern culture in which collective identity was so closely associated with the nation-state. Jews were perceived to be rootless and cosmopolitan, and they were subject to ongoing discrimination for that, preparing the way for their persecution in Nazi Germany and other nations of Eastern Europe. But in the second modernity, the Jewish heritage of diaspora was viewed more favorably in a world of mass migrations of both affluent elites who traveled freely and impoverished masses who sought to escape their miserable conditions. Now, Levy and Sznaider point out, Jewish cosmopolitanism had become closer to the norm of a globalizing culture in which allegiance to the nation-state was rivaled by alternatives. The symbolization of the Holocaust also eroded the sense of the exceptionalism of the Holocaust. As a marker of the extremes of genocide, it continued to provide the primary referent. But the politics of "ethnic cleansing" in Bosnia and Kosovo (which Levy and Sznaider label the "Kosovocaust") during the 1990s could not help but lead to comparisons of atrocities that were close in nature in their rationale and their execution.[57]

Particularly significant in the trend toward symbolization, Levy and Sznaider observe, was the Americanization of Holocaust memory. In American culture, universalization of the meaning of the Holocaust received its most far-reaching approbation for both political and cultural reasons. On the political plane, the USA led the way in declaring its preeminent role in the "new world order" of the post-Cold War era. The universalization of the motto "never again" had the dialectical effect of promoting an uncompromising imperative to champion human rights wherever

---

[56] Ibid., 17–19, 176–81.
[57] Ibid., 50–51, 156–62, 165–73.

they were violated.[58] On the cultural plane, Americans led the way in developing new technologies of media that would command the attention of mass audiences in pervasive ways. Levy and Sznaider enumerate significant media representations of the Holocaust that appealed to the ethical sensibilities of ordinary citizens by eliciting direct emotional responses to its horrors for innocent victims, ordinary people like themselves. These included the network television miniseries of the 1970s, the dramatization of Anne Frank's diary account of her hiding, the movie *Schindler's List* about a righteous rescuer, and Daniel Goldhagen's unsparing account of the evil deeds of the Nazis. In their ensemble, these media presentations dramatized the plight of victims with whom people of good will could readily identify and vicariously relate to their suffering. Herein, Levy and Sznaider claim, the publicity aroused through multimedia reconfigured historical remembrance in ways that circumvented the stalemate at which the professional historians had arrived, obviating their role as essential guides to Holocaust interpretation for the public at large.[59]

In the course of its redeployment as a symbolic icon demanding constant vigilance in the name of "never again," Holocaust remembrance was transformed into the signature of a future-oriented moral ideal, in contrast with the first modernity Holocaust memory that struggled interminably with the traumas of the past. But the universalization of such memory as a moral foundation for public policy, Levy and Sznaider observe, came with a price. First modernity memory so closely associated with the nation-state had demanded active civic obligation on the part of its citizenry, notably military service. Second modernity memory in its globalizing imagination evinced an exemplary display of compassion for strangers, but with fewer expectations of active involvement on the part of its citizenry. Its moral ideal was pervasive yet shallow as an ethics of responsibility.[60]

Second modernity memory, moreover, was conspicuous for its pluralism. In the twenty-first century, the globalization of Holocaust remembrance proceeded hand in hand with its localization. Deterritorialized in the process of its symbolic universalization, such memory was reterritorialized in particular places in different ways. Levy and Sznaider give three

---

[58] Ibid., 203–07. See the sequel by Levy and Sznaider, *Human Rights and Memory* (University Park: Pennsylvania State University Press, 2010) on campaigns for human rights.

[59] Levy/Sznaider, *Holocaust and Memory*, 36–38, 59–63,109–116–17, 132–43, 152–56, 162–65, 188–89.

[60] Ibid., 166–67, 173–79.

examples of variations: the republics of Israel, Germany, and the USA. On a spectrum from its first concrete formulations to its abstract universalism, each chose a different figuration of remembrance. Israel resisted the trend toward universalization. There, the Holocaust was officially remembered as a specific crime again Jews, a traumatizing event of overwhelming proportions in a long history of persecution of exilic Jews. Though prestate Zionists had done little to address the plight of European Jews in Nazi-dominated Europe, postwar Israel became a haven and defender of Jews everywhere. Germany, on the other hand, favored universalization because it lifted in some measure the burden of its guilt as a nation. In the USA, "never again" became the ethical foundation of American foreign policy. The terrorist attack on the USA on 9/11/2001, however, drew the American government into military intervention in the Middle East, setting for itself a role it has yet to master and that augurs interminable warfare against an elusive enemy into an uncertain future.[61]

[61] Ibid., 11–13, 83–95, 143–48.

CHAPTER 6

Nostalgia and the Mnemonics of Time

THE REVERSAL OF FORTUNE OF THE IDEA OF PROGRESS

The memory phenomenon opened a range of issues about the historians' understanding of time. French philosopher Jean-François Lyotard was to become famous among students of postmodern theory as the first to challenge the notion of a master narrative for modern history. His key to the postmodern temper of the 1980s was the repudiation of the idea that history might be plotted on a single timeline as the saga of the rise of Western civilization.[1] It was not just that Marxism as a philosophy of the progressive avant-garde was on the wane. Liberalism, too, with its mid-twentieth century commitment to the making of the welfare state through governmental responsibility for social planning, was falling back on the political attitudes of its nineteenth-century beginnings, which favored private initiatives and self-reliance. The age that gave wings to memory studies was also one that witnessed the revival of neo-conservatism.[2]

To dismiss the grand narrative as the essential timeline of modern historiography, however, was to open the way for an exploration of its mnemonic underpinnings, which embodied a future-oriented conception of historical time. The German historiographer Reinhart Koselleck formulated an interpretation of what he characterized as the "semantics of

[1] Jean-François Lyotard, *La Condition postmoderne; rapport sur le savoir* (Paris: Editions de Minuit, 1979), 11–17, 63.
[2] Tony Judt, *Ill Fares the Land* (New York: Penguin, 2010), 106–19.

© The Editor(s) (if applicable) and The Author(s) 2016
P.H. Hutton, *The Memory Phenomenon in Contemporary
Historical Writing*, DOI 10.1057/978-1-137-49466-5_6

time," with particular attention to the modern conception of historical time that emerged during the Enlightenment. He pointed out the great expectations of the historians of that era for the prospects of the future as a new age, and who therefore cast history as a saga of progress, with recognizable origins tending toward an anticipated future. The making of this idea was furthered by the birth of ideology at the end of the eighteenth century, with its programmatic schemes for the improvement of the human condition. The grammatical mode of such an understanding of history is the future perfect. History is written as if there were an expectant past preparing the way for its eventual fulfillment. In such a scheme, the present becomes a place marker in history's march toward its denouement.[3] To put this argument in more modest terms, a goal-oriented history cues the search for origins, and casts the present as but a stage along the way toward a foreseeable destiny. Koselleck further argued that modern historical consciousness betrays an ongoing tension between experience and expectation, or, alternatively, between memory and hope. Experience is conceived spatially in its references to places of memory; expectation, by contrast, is conceived temporally as a horizon of future possibilities. Koselleck proposed an inverse relationship between the two: the greater the expectation of the future, the more past experience contracts into a more precisely defined niche on the sequential timeline of modern history. Historical time is framed differently depending on the moment of time the historian privileges.

Koselleck's formulation was to have enormous influence upon historians reflecting on the issue of historical time from the perspective of our present circumstances. Whereas modern historiography favored the future as destination, some historians decided to apply Koselleck's formula in reverse mode. They proposed that we appoint the present as the privileged moment of historical time, a point of departure for looking back upon the past retrospectively. From this present-minded vantage point, the experience of the past loomed larger, while expectations for the future faded into uncertainty. Looking back on the past from the present dramatized its diversity and discontinuities. The dismal historical record of the twentieth century displaced the rhetoric of progress implicit in nineteenth-century forecasting, and it produced an assessment of many misgivings. It is in this context that the idea of nostalgia emerged as a perspective on

---

[3] Reinhart Koselleck, *Futures Past: On the Semantics of Historical Time*, trans. Keith Tribe (Cambridge: MIT Press, 1985), 246–88.

modern historiography in a search for lost time and missed opportunities. The scholarly interest in nostalgia was symptomatic of this changing perspective on historiography. Nostalgia concerned a world that had been lost, that had never measured up, or that was not but might have been. This reverse perspective on the idea of progress contrasted with the understanding of historical time sketched by Koselleck. Modern historiography had highlighted its expectations; studies of nostalgia redirected attention to its disappointments.

Scholarly Perspectives on the Reinvention of Nostalgia for the Modern Age

Scholarly interest in the topic of nostalgia came late to discussion of the workings of collective memory. But its moment may at last have arrived, bringing with it perspectives unappreciated a generation ago. As an emotional response to time's passages, nostalgia has long been viewed with suspicion. From the dawn of the modern age, critics have explained that it plays into life's illusions, drifting into sentimental idealization of a past on the fast-track to obsolescence. From the earliest critical commentaries on its nature in the late seventeenth century, nostalgia has been equated with homesickness, futile longing for lost places, lost times, and lost causes. For the most part, it was diagnosed as a psychological disorder that immobilized individuals susceptible to the tug of its emotions. It was in this guise that discussion of its nature entered the lexicon of medical discourse during the nineteenth century.

Nor did scholarship on the history of emotions, a somewhat earlier line of historical inquiry, find a place for nostalgia in its investigations. Nostalgia as longing for a golden age in the past may be as old as our species, as in ancient notions of a lost Eden. But the issue of changing ways in which nostalgia was experienced across the ages escaped the attention of such pioneers as Annales historian Lucien Febvre and his students in their research on the history of collective mentalities. Their interest was more in the emotional volatility that characterized the early modern period, a lingering effect of emotional life as experienced during the Middle Ages. For Robert Mandrou in his *Introduction à la France moderne* (1961), for example, love, anger, fear, and grief were common emotions that resided on the surface of the psyche, noteworthy for their intense and easily triggered instability.[4] Though he did not address the topic of nostalgia himself, German sociologist Norbert Elias's *The Civilizing Process* (1939), a study of the taming of emotions in the cult of manners during that era, set forth preconditions for the critical awareness of nostalgic emotions and the sensibilities they cultivated.[5] Feelings of regret once experienced naively gave way to sentimental reflection about them. So conceived, nostalgia at the threshold of the modern age might be regarded as the domestication of feelings about irretrievable loss in the face of changing social mores.

Thomas Dodman

Historian Thomas Dodman has studied the changing nature of the nostalgia of colonial exile, based on his research on psychological disorders among French soldiers and settlers in Algeria during the nineteenth-century occupation.[6] Military conquest of Algeria had been the easy task. The cultural adjustment of the forces of colonization proved more daunting. In the early days, he explains, physicians diagnosed the psychological distress of soldiers as homesickness, born of the difficulties of living in an arid landscape devoid of the comforting, familiar surroundings of home. Such nostalgia afflicted them in numbers that alarmed statesmen and generals overseeing the colonization. Some soldiers were sent to clinics. Tough cases were sent home. Pioneering settlers fared no better. As time wore on, what had been characterized as the malady of nostalgia was absorbed into a larger clinical diagnosis of maladjustment to an inhospitable tropical climate. Toward the end of the century, as the French presence became more deeply entrenched, the notion of nostalgia as pathology yielded place to one of tempered longing, a "sustainable nostalgia" in a land France was determined to make its own. Still, the project of maintaining French culture in Algeria was of such proportions that statesmen eventually launched projects to transplant entire French communities in hope of building within the colony a little bit of the world left behind. Dodman leads us toward a historical understanding of the ill-considered psychological and cultural costs the French paid to live in the land they had conquered.

[4] Robert Mandrou, *Introduction à la France moderne, 1500–1640* (Paris: Albin Michel, 1974), 75–89.

[5] Norbert Elias, *The Civilizing Process; The Development of Manners* (1939; New York: Urizen Books, 1978), 58–59.

[6] Thomas Dodman, "Un Pays pour la colonie: mourir de nostalgie en Algérie Française, 1830–1880," *Annales HSS* 66/3 (2011): 743–83.

He explains that the project of colonizing Algeria involved not so much familiarizing French settlers with unfamiliar surroundings as it did transporting what had been familiar into what would remain alien territory. Even after generations of settlement, he concludes, the French in Algeria were never completely at home.

Dodman's research on this clinical reductionism may explain why historians were slow to take up this topic in the now long-running scholarly discourse about collective memory.[7] But today, in the early twenty-first century, the workings of nostalgia, notably in their modern social and cultural settings, are receiving new scholarly attention, this time construed as an emotion that may be understood historically and collectively, not just psychologically and individually.[8] What is more, some scholars investigating nostalgia in the modern era—roughly the period from the late eighteenth to the mid-twentieth century—comment on its distinguishing traits, a kind of response to time's passages that for all its melancholy was self-revealing, reflective, even creative.

Fred Davis

Significant in this respect is the pioneering study by the American sociologist Fred Davis, *Yearning for Yesterday* (1979). Davis shows how the topic of nostalgia over the course of the nineteenth century escaped its once exclusive identification with medical discourse, and came to be appreciated as one emotion among many in the everyday lives of ordinary people. In time, the wistful sadness of nostalgia was acknowledged to be a normal acceptance of loss within the recognition of the irreversibility of historical change. Nostalgia, he argues, served as a safety valve for those

---

[7] It is true that scholarship on the history of commemoration, a central interest in memory studies dating from the late 1970s, provided an avenue toward understanding nostalgia as regret for a cherished past. But such studies emphasized the politics of contested identities, the interpretative strategy underpinning the notion of the "invented tradition." See Terence Ranger, "*The Invention of Tradition* Revisited," in *Legitimacy and the State in Africa*, ed. Terence Ranger and Megan Vaughan (London: Palgrave, 1993), 62–82.

[8] See the studies by Michael S. Roth on nostalgia as malady in nineteenth-century medical discourse: "Dying of the Past: Medical Studies of Nostalgia in Nineteenth-Century France," *History and Memory* 3 (1991): 5–29; "Remembering Forgetting: *Maladies de la Mémoire* in Nineteenth-Century France," *Representations* 26 (1989): 49–68; "The Time of Nostalgia: Medicine, History and Normality in 19th Century France," *Time and Society*.1 (1992: 281–84. See also Janelle L. Wilson, *Nostalgia; Sanctuary of Meaning* (Lewisberg,, PA: Bucknell University Press, 2005), 21–24.

who wanted to maintain a sense of continuity between past and present in their private memories, particularly as it became more difficult to do so in the public sphere. As he explains, nostalgia is a "crepuscular emotion," permitting the emotional survival of an idealized image of a past whose complex realities conflate as their memory begins to fade. [9]

Davis also remarks on the changing nature of modern nostalgia in light of the emergence of new technologies of communication that drew the imagined past more openly into the public sphere. Nostalgia may have been experienced privately. But it could be cued publicly by the image makers of journalism, advertising, and politics. Gradually but inexorably, these modes of representing the past at large shaped the way individuals integrated images of the past into their private recollections. Over the course of the modern era, he notes, this blending of personal experience and collective representation came to be so thoroughly interlaced as to be indistinguishable. Private memories were increasingly integrated into a common culture. To exemplify his argument, Davis alludes to the emergence of the notion of generational memory—as in memories of the cohort coming of age during the 1950s as opposed to that of the 1960s—shared reminiscences of signal events, or songs everyone loved.[10]

Peter Fritzsche

By the first decade of the twenty-first century, scholars were prepared to place Davis's notion of modern nostalgia in better developed historical contexts. Among the most important for the recent rehabilitation of the idea of nostalgia as it found expression during the modern era is that offered by European historian Peter Fritzsche. In a series of studies that culminated in his book, *Stranded in the Present* (2005), he proposes that nostalgia is an essential ingredient in the emergence of modern historical consciousness. His key point is that for the generation of Europeans coming of age in that era the felt experience of nostalgia as a response

---

[9] Fred Davis, *Yearning for Yesterday; A Sociology of Nostalgia* (New York: The Free Press, 1979), 1–29, 110. From a historiographical standpoint, Davis's discussion of the collective nature of nostalgia invites comparison with that of Maurice Halbwachs on the larger topic of collective memory. So too in their reception. At the time of their respective publications, their theories received little scholarly attention, only to come into play decades later. Like Halbwachs, Davis expounds on the relationship between social power and collective memory. Both explain how personal memories are localized within social contexts.

[10] Davis, *Yearning for Yesterday,* 111–16.

to rapid historical change may be interpreted as the reverse side of the ideologically driven discourse about progress. So conceived, nostalgia may be interpreted as a modern sensibility. Its emotions underpinned an awareness of the unpredictable, sometimes menacing realities of rapid and transforming historical change in an age whose public discourse favored rising expectations for the coming of a better world. [11]

Fritzsche, therefore, has explored this modern conception of historical time from the vantage point of its reverse mode—nostalgia for a lost past. Nostalgia, he contends, is a nineteenth-century invention. It makes manifest a growing awareness of the distance between past and present, and the need to savor the memory of a world that is fast disappearing and cannot be retrieved. Beginning with the French Revolution, he explains, precipitous change disrupted the lives of vast numbers of people, toppling long-established political regimes, driving social groups into exile, and in the process accentuating popular awareness of the widening divide between old and new ways of living. In the new world of rapid political, economic, and demographic upheaval, the experience of the past was no longer a reliable guide to present choices. Concomitantly, the accelerating pace of change led to unsettling anxieties about what the future might hold.[12] Ideas about time were being transformed, and in its midst nostalgia became the prevailing mode of memory. Fritzsche challenges scholars who dismiss nostalgia as a disabling melancholia to reconsider its complexity as an emotional response to life in turbulent times. While harboring the sadness of irreversible loss, memory in the guise of nostalgia can also quicken the resolve to deal creatively with an indeterminate future in which one's resources of hope may triumph over psychological resignation to irreparable loss. Nostalgia may sometimes have been more remedy than malady in the face of realities that denied the once reassuring constancy of tradition.[13]

---

[11] Peter Fritzsche, *Stranded in the Present; Modern Time and the Melancholy of History* (Cambridge: Harvard University Press, 2004), 45, 49, 142, 201–02, as well as his preliminary studies, "How Nostalgia Narrates Modernity," in *The Work of Memory: New Directions in the Study of German Society and Culture,* ed. Alon Confino and Peter Fritzsche (Urbana: University of Illinois Press, 2002), 64–65, and "Specters of History: On Nostalgia, Exile, and Modernity," *American Historical Review* 106 (December 2001), 1589, 1592.

[12] Fritzsche, *Stranded in the Present,* 11–54.

[13] Fritzsche, "How Nostalgia Narrates Modernity," 62–85; idem, "Specters of History: On Nostalgia, Exile, and Modernity," 1587–1618. See also Philippe Ariès, *Le Temps de l'Histoire*

Fritzsche's point of departure for his discussion of the modernity of nostalgia is the French Revolution, with its socially disruptive, life transforming consequences for people in all walks of life across Europe. Whatever the reforms promised and in some measure accomplished by statesmen sympathetic to the Revolution, its civil conflicts unleashed a reign of terror, the mass exodus of its opponents, uncertain often permanent exile for its hapless victims, together with the random death and destruction that were legacies of the Europe-wide wars of the Revolutionary and Napoleonic eras.[14]

As a historiographical point of reference with which to distinguish his own argument, Fritzsche invokes the model of German philosopher Reinhart Koselleck. His *Future's Past; The Semantics of Time* (1985) deals with the assumptions of the historical writing of the Enlightenment. Koselleck posits an inverse relationship between respect for experience out of the past and hope for the possibilities of the beckoning future. In light of the great expectations of the *philosophes* for social and political reform, the past of the *ancien régime* was discarded with few regrets. Fritzsche, by contrast, sets this relationship in reverse mode as he approaches the nineteenth century. As prospects for the future became uncertain amidst the turmoil of the Revolutionary era, the past waxed larger in a culture of retrospection on severed ties with a cherished world reduced to ruins forevermore.[15]

Fritzsche argues that such precipitous change aroused popular longing for the halcyon days of the *ancien régime*, at least as they were imagined in a deepening idealization of social life back then. The irony, he explains, is that such nostalgia was the seedbed of an emerging historical consciousness—by which he means not critical historical interpretation by scholars but rather an emotionally empowered recognition of the historicity of the human condition on the part of ordinary people. That stance on the past concerns the way singular events can and often do disrupt continuities between past and present, redirecting human affairs in unanticipated ways. The incentive to think historically about the human predicament, Fritzsche contends, is incited by these contingencies. Herein

(1954; Paris: Seuil, 1986), 33–43, who explains how his family's nostalgia for the traditions of old France served as his path into history.

[14] Fritzsche: *Stranded in the Present*, 11–54, 201–18; "How Nostalgia Narrates Modernity," 66–68; "Specters of History," 1594–1600.

[15] Fritzsche:"How Nostalgia Narrates Modernity," 67, 76–77; "Specters of History," 1589–91, 1602.

ordinary people became personally aware of larger historical forces at play, as change beyond their control forever altered their private lives. The rapid succession of life transforming events conveyed a sense that time was accelerating.[16] Fritzsche's argument conveys an irony. Nostalgia, once judged a psychological malady, was reconceived as an emotion that sensitized exiled or displaced people to an understanding of the realities of historical change. Nostalgia narrated their experience of the near past as stories told about changing fortunes, adventures, and migrations in the midst of turmoil. Such narratives became the folklore of the nineteenth century, recompense for those buffeted by unwelcome forces of history that were redirecting the course of their lives.[17]

It is interesting to juxtapose Fritzsche's interpretation of the nostalgic nineteenth century to the one advanced by left-wing French historians of the Revolution, from Jules Michelet to Michel Vovelle. Their interpretation was informed by progressive ideologies born of the Revolution, and at a deeper remove, by the expectant assumptions of the *philosophes* of the Enlightenment about what the future held. Fritzsche offers an alternative way of framing nineteenth-century history. History does not follow an anticipated pattern based on human projects for reform. Nor does it reify a "direction of moral intention," as Georges Lefebvre, dean of historians of the French Revolution, once interpreted the course of modern French history. Like the historians, the Revolution's statesmen may have forecast the future of history, but their perceptions were rarely congruent with those rising out of the lived experience of ordinary people coping with their newfound situation.

Catastrophes in Europe on such a scale may predate the French Revolution. One thinks of the Thirty Years War of the seventeenth century as an example of life-changing havoc across Europe, to be accepted with fateful resignation. By the early nineteenth century, however, those buffeted by precipitous misfortune sought compensation as never before in private worlds of sentimental consolation. In such circumstances, the

---

[16] Fritzsche, *Stranded in the Present*, 7, 77, 88,154,159. Here, we might say, Fritzsche introduces a historiographical interest in the tempo of historical time that contrasts dramatically with that of the once popular Annales school of historiography, which placed its accent on the inertial pace of time, that is, time as immobilized in the preindustrial societies of early modern Europe. See Emmanuel Le Roy Ladurie's essay, "L'Histoire immobile" in his *Le Territoire de l'historien* (Gallimard,1978), 2: 7–34.

[17] Fritzsche: *Stranded in the Present*, 80–83, 87, 90, 128; "Specters of History," 1607–1609.

modern bourgeois family became a sanctuary of nostalgia. The lost past that people remembered was idealized as heritage and nested within the domestic interiors of their households. Emotions nurtured in the midst of family intimacy waxed large in their lives. The interior of homes became places for cultivating personal memory. People collected souvenirs with a sense of purpose and furnished their homes with such memorabilia. The bourgeoisie, fashioners of a new urban and industrial culture in the public sphere, cultivated personal memories of the private one closer to their everyday lives, as portrayed in their memoirs, autobiographies, diaries, and letters. In these myriad of ways, they deepened what might be characterized as the privatization of memory.[18]

Charles Rearick

As the century progressed, feelings of nostalgia settled into a more wistful mode. In his book *Paris Dreams, Paris Memories* (2011), historian Charles Rearick analyzes what is typically thought of as mainstream nostalgia—bittersweet memory domesticated in a sentimental journey into a much-loved past.[19] Such a conception of nostalgia might be situated historically between the deep nostalgia of psychological distress (the first mode in which it was interpreted critically) and the shallow nostalgia of consumerist kitsch that reigns today. Such was the memory of the everyday culture of Paris circa 1900, as idealized after World War I by Parisians as *la Belle Epoque*. Rearick's key point is that nostalgia for the everyday culture of the city in that era is conspicuous for its staying power, cherished as a core of memories in which Parisians found comfort and consolation through all the turbulent events of the twentieth century. The notion conferred upon Paris a cultural identity that has endured to this day. Paris is a city of many-layered historical identities. But the concentrated image of Paris in those years before World War I, Rearick argues, has outshone all others in its nostalgic effect. He sets out to explain why.

In tracing the emergence of modern nostalgia for Paris in the *Belle Epoque*, Rearick notes a close association between war and remembrance. Insofar as everyday life in Paris circa 1900 had been pleasant, it had been

---

[18] Fritzsche: *Stranded in the Present*, 9, 79, 160–200; "Specters of History," 1600–01, 1605, 1616; "How Nostalgia Narrates Modernity," 79.

[19] Charles Rearick, *Paris Dreams, Paris Memories; The City and its Mystique* (Stanford, CA: Stanford University Press, 2011), 44–81.

appreciated in an unreflective way. But after World War I, and echoing again after World War II, Parisians self-consciously yearned for what they remembered as the sweet tranquility and well-being of life back then before these wars. The term *Belle Epoque*, he explains, did not come into common usage until the mid-1920s. It was a memento of what was perceived to be a vanishing way of life that ordinary Parisians idealized, and to which they referred affectionately as *Paname*. Such a life was remembered for its harmonious social relationships in an era of tranquility, as if it had been unmarred by social strife. Everyday life in such sentimental recollection was carried out on a human scale in what had been thought of as the "villages" of the city. During the interwar years, appreciation of that era was cast in aesthetic images in lyrical song in the cabarets, the popular entertainments of its dancehalls, and the camaraderie of its bistros and cafés. Popular nostalgia for *la Belle Epoque* was given a boost by writers and artists who portrayed its appealing charms for the public at large, and eventually for tourists from abroad who would one day flock to the city to experience directly if only fleetingly its legendary mystique. The publication of guidebooks to the sites of "old Paris" further enhanced this publicity, whose gathering force prompted the municipal government of Paris to throw its support behind projects for preservation of its landmarks, with an eye to the economic benefits of these initiatives. There were, of course, critics of the nostalgic take on late nineteenth-century Paris, who sought to puncture bucolic imagery by testing it against the tough social realities of the era that had at the time contributed to social discontent—the social rifts engendered by the Dreyfus affair with its anti-Semitism and incitement to violence, the dull sameness of projects for modernization of the built environment, the disappearance of rural vestiges in the city. But even in their aggregation, Rearick notes, these historical realities were not imposing enough to dispel the myth of life in those times as Parisians wished to remember it. More than any city in the Western world, he contends, the modern identity of Paris came to be identified with the cultural memory of a bygone era.

Rearick cites two issues that stand out in the making of the nostalgic image of *la Belle Epoque* for posterity. The first was the horror of World War I and the privations of its immediate aftermath. Amidst the sadness of lives lost, dearth of foodstuffs and other provisions, and deprivation of the amenities of life, idyllic images of a kind of life swept away by war loomed large in popular remembrance. The second factor was the long-range and ongoing process of modernizing the urban landscape of Paris. The quaint charm of the built environment of "old Paris" gradually but relentlessly succumbed to the drab sameness of newly fabricated apartment buildings and commercial establishments, constructed for efficiency and devoid of aesthetic appeal. As the map of Paris was over time transformed into a modern cityscape, the tangible places of *la Belle Epoque* began to disappear. A few architectural icons survived, and received all the more nostalgic attention for that. As a place of memory, Montmartre was the most prominent among them. Citizens of that quarter displayed resilience from the late 1920s in dedicating themselves to preserving what remained of their heritage. Montmartre became a center of resistance to large-scale building, the first stirrings of urban environmentalism. Associations formed to fight developers and occasionally they prevailed. In a way, *la Belle Epoque* was reinvented in that self-contained space by artists, filmmakers, and writers who portrayed its past as an imagined community of picturesque appeal for its denizens and tourists alike.

The remaking of nostalgia for the *Belle Epoque* after World War II, Rearick explains, echoed the pattern of that after World War I. Following a few years of hardship, the economy of Paris began to rebound and collective nostalgia flourished once more, as theme-based simulations of life back then in the theaters and cabaret reviews were revived to entertain a burgeoning tourist presence. By that time the notion of a *Belle Epoque* had expanded to include the interwar years, even as more of its memorable places disappeared. So reconceived, Rearick argues, the remembrance of Paris circa 1900 continued to shape the cultural identity of Paris into the 1970s. Even after that date, he notes, reminiscence about the *Belle Epoque* popped up now and then, as in Woody Allen's Paris nostalgia film, *Midnight in Paris* (2010). By then, however, Paris circa 1900 was too remote for anyone to remember it with living evocations of feelings of regret. The *Belle Epoque* had become "retro"—a world into which one might endeavor to enter imaginatively but could no longer do so with the tug of reality. By then, nostalgia itself had passed into its postmodern, consumerist mode—kitsch stripped of yearning for a lost time in the life of the city.

Considered in larger terms, Rearick concludes, the imagined community of *la Belle Epoque* draws attention to the ongoing quarrel between preservationists and modernizers that has grown more confrontational in times closer to our own. Relentlessly, the modernizers might appear to be winning, in Paris as in every major city of the world. Still, Paris, more than any of these, has managed to preserve just enough places of tangible memory to sustain its allure as a beautiful city of dreams. Paris is exemplary, Rearick argues, for its dedication to its heritage as it has been layered century upon century, but especially for that period around 1900, a lasting memento of an imagined *douceur de vie* to which posterity has repeatedly turned amidst all the disasters that have befallen the city over the course of its twentieth-century history.

Svetlana Boym

Taking the concept of nostalgia in a different direction, noteworthy for our review of its historical dimensions is the influential study by the Russian-American critic Svetlana Boym (1966–2015), who takes up the theme of nostalgia's reflective side in *The Future of Nostalgia* (2001). Whereas Fritzsche writes of nostalgia born of historical contingencies, and Rearick of an enchanted past, Boym recasts the idea of nostalgia in light of worthy yet missed opportunities for human betterment recovered from a discarded past. A voluntary exile from her native Russia when it was still the graveyard of Bolshevik dreams, she immigrated to the USA, where she became a novelist and professor of comparative literature at Harvard. Yet for all her success in making a fulfilling life in the new world, her heart remained engaged with the one she had left behind. Despite its intellectual complexities and high literary motifs, her *Future of Nostalgia* is a rather personal book, giving expression to her ambiguous feelings about her relationship with her native land.[20] One among that cohort of the Russian intelligentsia that departed for foreign shores during the 1970s and 1980s, her interest in nostalgia was born of her exile, now examined critically for its positive as well as its negative effects. The workings of nostalgia, like other expressions of collective memory, she explains, are caught up in a dynamic process of remodeling, sometimes remaking the old world creatively in the new, localizing its unrealized yearnings amidst present realities. "One is nostalgic not for the past the way it was," she remarks, "but for the past the way it could have been. It is the past perfect that one strives to realize in the future."[21] Her interest in nostalgia is less about loss, more about reinvigoration. In light of the collapse of the Soviet Union as a failed experiment in the making of the good society, she investigates

---

[20] Svetlana Boym, *The Future of Nostalgia* (New York: Basic Books, 2002), xiv–xv; www.svetlanaboym.com
[21] Ibid., 351.

the resurfacing of discarded visions out of the past about what the future might hold. In other words, her interest is not in elegy for a world that we have lost but rather a reverie for one that might have been. Transporting past dreams of the future once denied into a more appreciative present, nostalgia takes on a utopian allure.[22]

Like Fritzsche, Boym's point of departure is a political revolution—in her case that of 1989 in Eastern Europe, prelude to the collapse of the Soviet Union 2 years later, dying anyway under senescent leadership resistant to the technological innovations of the media revolution. But it was not just the demise of the old regime that intrigued her. It was the particular experience of Russia between her twentieth-century revolutions. The Soviet Union was born of a vision of bringing into being an egalitarian society. The vision was dispelled amidst the realities of ongoing Bolshevik rigidity and oppression during its 70-year history. But conceptions of what that revolution might have been coalesced from time to time in resistance movements conjured up along the way. Bolshevism may have destroyed or intimidated all of them. But the demise of the Soviet regime in 1991 opened for examination the memorable remains of alternative conceptions of what the good Russian society might have been. The imaginary landscape of the erstwhile Soviet Union was littered with discarded architectural and literary artifacts, now open for reinterpretation. Boym went in search of them as mementos of lost causes, to be found in such places as the theme parks of Moscow and St. Petersburg, as well as in the domestic interiors of Russian exiles in America.[23]

Boym takes seriously the proposition that there need be no contradiction between emotional longing and critical thinking. She makes her case by formulating a distinction between two kinds of nostalgia, restorative and reflective. Restorative nostalgia is the one with which we are more familiar. Those captured by its emotions long for the perceived simplicity and harmony of an earlier age with which they are determined to maintain continuity. They want to keep that past "forever young" and so give themselves to illusions about the way it was. Reflective nostalgia, by contrast, accepts the disruptions, ambiguities, and complexities of the past as well as its place on an irreversible timeline. For Boym, it is ironical, inconclusive, and fragmentary. It seeks not a cure for its discontents, but rather meditation on the possibilities of dealing with its disappointments

---

[22] Ibid., 57–71.
[23] Ibid., 83–91, 327–36.

by refashioning old dreams in new settings. Reflective nostalgia values the past for its futurist speculations—dissident ideas discarded along the way, but now retrieved to be woven into the fabric of a different time in history. She treats such reflections as cross-grains in Russia's cultural heritage during the Soviet era. This latter notion was her more original formulation of the way in which nostalgia might be conceived.[24]

In revisiting the Soviet experience from her twenty-first-century vantage point, Boym recounts how both kinds of nostalgia were operative during the early days of the post-Soviet era, for life in the Russian Federation of the 1990s never measured up to expectations. For some there was a longing for the material and psychological security of life under the Soviet regime. But others preferred to remember the resistance movements of that regime's later years. Ironically, the counter-cultural liberation movements that emerged to challenge the Soviet regime (much as did like societies in Czechoslovakia, Poland, and Hungary) lost their momentum, and one might say the immediacy of their meaning, once the revolutions of 1989 had taken place. Reflective nostalgia, therefore, was a way to revisit and assess the meaning of the dreamwork that lay behind the projects of oppositional movements during the Soviet era. Over time they had become more revealing about subtle cultural defiance than of openly subversive opposition.[25]

POSTMODERN NOSTALGIA

Conceptions of nostalgia have evolved with the times, giving rise to what might be characterized as postmodern nostalgia. These treat the past less reverently. In a noteworthy article that appeared in 2005, University of Toronto professor Linda Hutcheon reflects on the irony implicit in today's expressions of nostalgia. Nostalgia in our times, she explains, must be understood in terms of its interplay with irony, a perspective that puts experience of the past at a critical distance while simultaneously integrating its feelings within the imagination of the present in inventive ways. She allows that she had just published a book about postmodern irony, only to note—not without irony—that nostalgia was being drawn into its sphere of influence. She remarks upon a historical transition in the understanding of this emotion, for her critical perspective complicates and departs

---

[24] Ibid., 41–55; "Nostalgia and its Discontents," *Hedgehog Review* 9 (2007): 7–18.
[25] Boym, *Future of Nostalgia*, 61, 149–56.

from assessment of its nature in earlier times and places. She discusses the implications of the shift, arguing that nostalgia in our times has been reconceived to suit a postmodern mentality, one conveying the ambiguity of a present mindedness uncertain about its relationship to both past and future. Both irony and nostalgia are literary tropes, she explains, the former dispelling sentimentality, the latter luxuriating in it. As a subjective response to life's fortunes, neither perspective lives in the experience of the past, but rather inheres in the mind of the observer seeking to appreciate it. One woman's irony may be another's nostalgia. Much depended on the meaning one invests in the past, she notes, and that is a matter of personal expectation. Still, she concludes, any expression of nostalgia in our present-minded times is likely to be tinged with irony.[26]

Fred Davis, too, sketches an interpretation of this postmodern remodeling of collective nostalgia in our contemporary age, as these emotions have come under the sway of the ever more imposing power of mass media to take control of the public representation of the past. The agency of this change, he argues, is postmodern consumerism. Increasingly, media in the interest of commercial profit appropriates and redeploys images of the past calculated to inspire collective feelings of nostalgia. Nostalgia in the contemporary age has fallen prey to the wiles of advertising, upsetting the earlier balance between its public and private expression. It has been incorporated into a consumerist culture, in which private emotions are manipulated for commercial advantage with the willing participation of its clients. Here it is not the individual idealizing the experience of the past but rather the image makers of Madison Avenue simplifying and sanitizing its representation in ways that promote a self-indulgent culture of materialist desire.[27]

Davis's analysis of contemporary nostalgia resonates with that of literary critic Fredric Jameson, who addressed the topic at roughly the same time. Jameson, instrumental in defining the concept of the postmodern in the late twentieth century, coined the provocative notion of "nostalgia for the present," a perspective emanating from the paradoxes of late capitalism in its quest to promote an economy of desire over one of need. Late capitalism redirects attention to consumption, and nostalgia has been

<hr>

[26] Linda Hutcheon, "Irony, Nostalgia, and the Postmodern," (University of Toronto English Department, 1998) http://www.library.utoronto.ca/utel/criticism/hutchinp.html

[27] Davis, *Yearning for Yesterday*, 118–42. See also David Gross, *The Past in Ruins* (Amherst: University of Massachusetts Press, 1992), 75–76.

pressed into the service of commerce as a marketing tool. One fabricates fantasies of the past with which today's consumers can be enticed into vicarious emotional identification. Jameson references novelists and film-makers of the 1980s who reinvigorated a fading memory of America of the 1950s by setting forth in high relief alluring yet distorting images of lifestyles back then. The effect is to reduce the cultural complexities of the era to extravagantly manufactured stereotypes. "Eisenhower's America," he allows, is reproduced as a "Potemkin Village" to satisfy the audience's wish to reenchant that world. For Jameson, the mode of postmodern nostalgia is irony, considered less as critique more as bemused detachment. In this guise, it expresses longing not for the past that was but paradoxically for one recast in imagery that satisfies present-minded consumerist expectations. Nostalgia in this guise is "retro," more appealing for its aesthetic gloss than for deeper currents of emotions that coursed through nostalgic imagery a century ago.[28]

Reaction to postmodern nostalgia so conceived—manipulative in its present mindedness and shallow in its ironical display—may in some measure account for the historians' newfound interest in the nostalgia of the modern era as a distinctive state of mind, a longing to hold on to a conception of the past that, for all its idealization, still managed to convey the authenticity of feelings that issued from the depths of time. One might argue that the transit of nostalgia from immobilizing homesickness to wistful remembrance signifies a taming of emotions, much in the manner that the sociologist Norbert Elias explained the binding of emotions to social conventions in the civilizing process over the course of the modern era.[29]

Scholarly interest in the idea of nostalgia, particularly as conceived in our times as regret over lost opportunities in the past, was the first response to the crisis of late twentieth-century historiography. It remained situated within the mnemonic framework of the linear conception of historical time outlined by Reinhart Koselleck, albeit in reverse mode of looking back upon the failure of the modern age to fulfill the expectations of the Enlightenment for the future of history. An interest in presentism as an alternative formulation of the mnemonics of time soon followed. The term is a neologism. By the late 1980s, however, it had become common

[28] Fredric Jameson, *Postmodernism, or the Logic of Late Capitalism* (Durham, NC: Duke University Press, 1991), 19–20, 156; "Nostalgia for the Present," *South Atlantic Quarterly* 88 (1989): 517–37.

[29] Elias, *Civilizing Process*, 58–59, 134–43, 180–205.

currency in the vocabulary of historians. The concept, with its implications for revaluing the uses of collective memory, was embedded in a discourse about postmodernism.

WALTER BENJAMIN: NOSTALGIA AND HISTORIOGRAPHY IN OUR TIMES

I close this chapter with a brief reference to the German literary critic Walter Benjamin (1892–1940) for the ties he envisioned between nostalgia and historical understanding. Coming of age in the years between the World Wars I and II, he set the tone for scholars in our times who value his insights into the relationship between memory and history. He was to become a memory figure himself during the 1990s, and played a role as intellectual celebrity not unlike that of Michel Foucault during the 1980s.

Benjamin was a brilliant, if eccentric student during his youth, later an earnest and relentlessly intellectual man of letters, admired by friends but little known outside of the literary circles for which he wrote during the interwar years. He found refuge in Paris after Hitler came to power in Germany in 1933. He frequented the Bibliothèque Nationale and lived on the edge of poverty with modest subventions for his essays of literary criticism. He continued his literary studies through the 1930s, but was increasingly drawn to ideas about history in light of the troubles of his times. His essays on historiography have a nostalgic allure, for he looked to the past rather than the future for insight into how Europe in the present age might escape Nazi oppression. He chastised the German Social Democrats with whom he was sympathetically allied for their fainthearted capitulation in the face of Hitler's intimidating assault on the Weimar Republic. On a more intellectual plane, he repudiated the idea of progress in history as an empty notion.[30] Historians of our own times intrigued with his ideas have been especially taken with his essay "On the Concept of History" (1940), hastily written on the eve of his flight from Paris as German military forces approached the city. Benjamin died a few months later on the French/Spanish frontier. A commemorative monument to his memory has been erected there, and a certain nostalgia for the broken promise of his life has persisted among his colleagues and sympathetic readers to this day. Benjamin's longing for what might have been appealed

---

[30] Patrick Hutton, "Walter Benjamin: The Consolation of History in a Paris Exile," *Historical Reflections* 36/1 (2010): 76–94.

to their own reflections on the failed promise of modern history for the making of a better world. In his poetical way, he prefigured the turn of postmodern historiography from the prospects of the future toward the redemption of the past.

Benjamin composed his essay as a series of aphorisms in the manner of Karl Marx's "Theses on Feuerbach" a 100 years before. Though he admired Marx and was drawn to his ideas, he repudiated his historical determinism and placed his own hopes in historical contingencies in which memory exercises its inspiring powers. The guiding motif of his essay was the Angel of History, an image taken from a painting by Paul Klee that was to become the subject of his meditation. The Angel looked back in sadness upon the debris of the failed projects and dashed hopes of the modern historical era, all carried out in the name of a progress that turned out to be an illusion. But he culminated his essay with reference to his own mildly messianic faith that humankind would find its way to renewal through the "profane illumination" that memory can provide. Here Benjamin put his accent on significant conjuncture rather than long-range historical developments, and he spurned the notion of historical inevitability. He pointed to memory's consolations in the midst of the seeming impasse of present circumstances, with a sense of kinship with people out of the past who shared his humane values. Benjamin never abandoned his hopes for the future. But his nostalgia for that future lay in his faith in the eternal return to memory's rejuvenating powers to show the way. Such nostalgia for what might have been, and the faith that it might yet be made real, suggests why Benjamin has attracted so much attention among today's historians.[31]

---

[31] Walter Benjamin, "On the Concept of History," in *Walter Benjamin: Selected Writings*, ed. by Howard Eiland and Michael Jennings (Cambridge, MA: Harvard University Press, 2003), 4: 389–400.

CHAPTER 7

Negotiating the Boundary between Representation and Experience

Memory's Newfound Claims Upon History

In its many venues, the intense interest in the history of memory had the unintended effect of unsettling long-established conventions of historical narration. Historians justifiably profess impartiality and dispassion in their research and writing. But memory studies called attention to subjective factors in historical interpretation that challenge their claims to objectivity. History and memory may be of a different order. But as the ongoing discussions of the relationship between them have revealed, they impinge on one another in ways that blur the boundary between them. Historiographical discussion of the memory phenomenon could not help but highlight the mnemonic character of historical interpretation. Phrased in the parlance of memory studies, historians provide mnemonic cues to their readers in the way they write history. Historians have the power to frame what the public recollects out of the past. If historians are the guardians of public memory, they are its arbiters as well. They not only sanction the past that is to be remembered but also shape the way it is presented. As a minimum, memory studies led some historians to suggest that problems of interpretation be addressed with greater modesty by acknowledging the realities of bias, pyschoanalytical factors in authorship, and the limits of historical representation. As a maximum, such studies raised broader issues about strategies historians employ to move closer to the subjective experience of historical actors out of the past. Nearly all the early work in the historiography

© The Editor(s) (if applicable) and The Author(s) 2016
P.H. Hutton, *The Memory Phenomenon in Contemporary Historical Writing*, DOI 10.1057/978-1-137-49466-5_7

of memory studies focused on issues of historical representation. But as historians explored its possibilities, some came to reflect on its counterpoint—experience as representation's existential ground. If there are limits to historical representation, they asked, may the boundary between them be pressed to draw us closer to the past as it was experienced by its historical actors?

The changing tenor of historiographical discussion of the issue of historical objectivity is evident in Peter Novick's *That Noble Dream* (1986), widely adopted as a basic text in graduate historiography courses in American universities.[1] Challenging the "noble dream" of historical objectivity, Novick sought to expose the bias, distortions, and omissions in the master narratives of American history. He pointed out how American historians with a certain naiveté had long presented a past they wanted to remember. From the founding of the American Historical Association in 1884 until well into the twentieth century, eminent historians tended to favor a patriotic view of American identity that denied the divisive realities of class conflict, racial and ethnic discrimination, and the diverse viewpoints of an expanding immigrant population. He drew attention to the near impossibility of obtaining complete professed detachment, and to the insidious temptation to treat objectivity as if it were no more than a consensus of viewpoint promoting professional harmony. The noble dream of historical objectivity, he maintained, is an elusive quest when tested against the actual debates about the past that have impassioned American historiography since its inception. As this historiography of patriotic consensus fragmented from the mid-twentieth century, he pointed out, a new generation of practicing historians sought to reclaim the forgotten past of women, African Americans, Native Americans, and other marginalized groups, while those with a theoretical bent proposed new categories of conceptualization to frame a more complex historical memory, notably through models for gender studies, the history of collective mentalities, and global history. In the process, they subverted the political identities previously highlighted by modern American historiography. Implicit in his presentation of historiographical controversies from across American history is the notion that it is better to understand and accept contested interpretations than to deny them in the name of a specious objectivity.

---

[1] Peter Novick, *That Noble Dream; The 'Objectivity Question' and the American Historical Profession* (Cambridge: Cambridge University Press, 1988).

In problematizing history's subject matter, memory studies have contributed to the widening interest in historiography since the 1980s. Once a technical subject dealing with methods for laboring in the archives, historiography has been reborn as a study of the conceptual schemes in which history is framed. Put differently, memory studies have played a role in the shift from a preoccupation with problems of evidence in historical research to those of rhetoric in historical writing. Historiography, once focused on issues about finding and evaluating sources, has been reoriented toward those of strategies for plotting narrative. Historiography today, therefore, operates at a far remove from Jacques Barzun's *The Modern Researcher* (1957), the essential primer for historiography courses during the 1960s, for the memory phenomenon raised new issues about the interplay between memory and history and so permitted historiography to assume center stage.[2]

In this chapter, I consider the relationship between the memory phenomenon and postmodern theory, with particular attention to the foundational studies by Keith Jenkins and Frank Ankersmit. The French philosopher Michel Foucault serves as an intermediary between them not only for his scholarship but also for the way in which he himself has been remembered. Foucault insisted in a radical way on the power of the form as opposed to the content of public discourse. But upon his death, historians immediately sought to decode the secrets of his own subjective personality. The interest in the subjective Foucault was symptomatic of the reconsideration of the relationship between representation and experience. Scholars stressing representation maintain a sober critical distance from their subject matter, emphasizing limits upon the modes of portraying subjective experience. Those favoring experience, by contrast, inquire into the possibilities of vicarious emotional identification with historical actors of the past, seeking to make them "come alive again" through such performative modes as tourism to historical sites, historical drama, and the historical reenactment movement. To illustrate my argument, I review the work of Ann Rigney on historical fiction. I close with a discussion of the interpretation of historian François Hartog, who places today's presentist stance on historical time in historiographical perspective.

---

[2] Jacques Barzun and Henry F. Graff, *The Modern Researcher* (1957; Florence, Kentucky: Thomson/Wadsworth, 2004) is now in its sixth edition.

KEITH JENKINS ON POSTMODERN HISTORIOGRAPHY

The scholarly discourse on postmodernism ran parallel with that on memory through the 1980s, though postmodernism's impact on scholarly research initially received more scholarly attention. Postmodernism was a rejuvenating intellectual venture of the 1970s. Insofar as it may be identified with historical scholarship, it was launched by Hayden White (b 1928), who directed attention to the role of rhetoric in historical composition. Historical writing draws on a repertoire of styles of composition. There is a poetics to the figuration of historical narrative. The interest in historical representation, therefore, dominated discussion of historiography during the 1970s.[3] I argue that the rhetorical turn of scholarship in the 1970s invited the mnemonic response during the 1980s. Whereas postmodern historiography put its accent on modes of representation, history written with the protocols of living memory in mind emphasized experience as a counterpoint. How the boundary between representation and experience has been negotiated since then is the subject of this chapter.

British scholar Keith Jenkins (b 1943) draws forth the implications of postmodern theorizing for rethinking the relationship between memory and history. In addition to his own writings, he has edited a reader on postmodern historiography, in which he includes 37 authors who have contributed in varied ways to our understanding of the concept.[4] Some are historians, others philosophers, still others literary critics. His talent has been to distill the essential elements of what such a historiography might be. Given the multitude of incongruent interpretations of the nature of the postmodern condition, Jenkins applies its precepts to historical understanding with disarming simplicity. His unifying theme is the dismantling of history from its standing as a science, assigning it a more modest place among the arts of memory.

Jenkins builds his argument as a polemic against positivist historiography. His thesis might be construed as a variation on the notion of the "end of history" in our times, understood as a way of thinking about history. Most professional historians, he contends, cling to a conception of historical scholarship that dates from the late nineteenth century. Scholars working in this tradition view their methodology as a science, based on the

---

[3] Hayden V. White, *Metahistory; The Historical Imagination in Nineteenth-Century Europe* (Baltimore: Johns Hopkins University Press, 1973, and his more recent *Figural Realism; Studies in the Mimesis Effect* (Baltimore: Johns Hopkins University Press, 1999).

[4] Keith Jenkins, ed., *The Postmodern History Reader* (Oxford, UK: Routledge, 1997).

historicist assumption that there is a backbone to history that the historian can discover and ultimately describe in an overarching grand narrative. They base their authority on their ability to write about that past with objectivity, dispassion, ideological neutrality, and careful documentation of their research. They profess to respect the past as past, worthy of study for its own sake. They believe that they can access its realities directly and come to understand them comprehensively in a culminating interpretation that synthesizes research and builds toward a consensus that approaches the truth about the past. As such, they see themselves as indispensable intercessors between past and present for the public at large.[5] Jenkins judges historiography so conceived as a failed venture, without credibility as a theory of knowledge, remote from the needs of the present age, and today in disarray as a working model for historical scholarship. From his perspective, the historians' quest is a search not for truth (as understood in the natural sciences as predictable patterns) but rather for meaning as taught by scholars in the humanities, which may vary from age to age but for whom the common goal is wisdom about how to live.[6]

Jenkins's polemic against positivist history enables him to cast his vision of a postmodern alternative in bold relief. His argument goes something like this: Historical inquiry should proceed from a present-centered perspective. The past was real but is no longer so, in the sense that historians in the present cannot touch those realities directly, but only represent them metaphorically. In characterizing that past from a presentist perspective, he prefers the term "before now." The claim that the analysis of data, the keystone of modern historiography, is a truth game is based on a misplaced notion of concrete certainties embedded in facticity. Facts about the past do not stand alone as autonomous entities, but rather acquire meaning only as they are contextualized within narratives. It is as a story that we experience life, and history provides a perspective that lends meaning to that experience. The story of history resists closure of the sort that positivist historiography would impose. Historical scholarship proceeds dialectically through revision of standing arguments, as each generation of historians revisits the past anew. Far from fixing the past for the present, historical interpretation moves with the times. From a postmodern perspective, historical interpretation is a

---

[5] Keith Jenkins, *Refiguring History; New Thoughts on an Old Discipline* (London: Routledge, 2003), 2, 9, 39, 59.
[6] Ibid., 5, 39.

dynamic, never-ending story. The pageant it portrays is endlessly revised as historians reconsider the meaning of the past in the eternal present. Historical inquiry, therefore, is better understood not as a search for some objective truth about the past but rather an interpretation of its meaning in the now time in history—meaning that we can appreciate in light of a correlation between our experience now and that of historical actors back then. Historical interpretation in its postmodern guise is creative and fictive, an imaginative reconstruction that represents what life was like in the past. In this, history is akin to that of literature and other genres of aesthetic appreciation in its striving for verisimilitude. Accordingly, history as a mode of intellectual inquiry is informed by the arts of rhetoric rather than the methods of the natural sciences. The key to historical understanding lies in the style of its figurations, that is, the way in which authors fashion their narratives. The success of these initiatives depend upon the plausibility with which they are able to render the meaning of human experience back then (the "before now") in light of what we know of experience in our own lives.[7]

Jenkins's thesis has far-reaching implications for understanding the relationship between memory and history. Modern historiography drew a sharp line between them. Postmodern historiography draws them closer together in a way that makes them almost interchangeable. Postmodern history, like oral tradition, is present-minded, restless to move on, agonistic in the play of interpretation it inspires, repetitious in its acts of interpretation while endlessly reconfiguring them. By immersing oneself in the act of imagining of the past, one strives to overcome the distance between past and present by time traveling between the two.

Jenkins might be accused of formulating a reductionist version of modern historiography that equates its practice with a naive nineteenth-century positivism long since superseded.[8] It is questionable whether any professional historian today takes seriously the position he challenges. He might be criticized as well for making light of the hermeneutics of interpreting the past. In the hermeneutical encounter between past and present, the historian is advised to be modest about the limits of understanding that our present experience permits as we enter into worlds whose experience does not accord with our own. One value of historical inquiry is to con

---

[7] Ibid., 3–8, 27, 39–58.
[8] See Michael S. Roth, "Classic Postmodernism," *History and Theory* 43/3 (October 2004): 372–78.

front the past in its strangeness, to be appreciated for introducing us to experience with which we may have little or no familiarity. In this respect, his approach does not address experience at the limits of human comprehension, of the sort on which Saul Friedländer ruminated in his discussion of the sublime evil of the Final Solution. In his focus on the composition of narrative, moreover, Jenkins has nothing to say about the weight of evidence underpinning it, or of unresolved burdens issuing from the past with which we have yet to reckon. One wonders, too, about the proposition that the past is not real. It might better be framed as one that takes into consideration epistemological limitations about what may be recalled with certainty about the past in the present. Jenkins equates truth with certainty, and there is evidence of the past about which we in the present can be certain. Much of the work of the historians concerns establishing such facts. Consider, for example, recent work on the migration of humans deep in the past, based on DNA genetic analysis. It provides a record of genetic inheritance and hence of the movement of peoples about which we can be certain.[9] What the record means as it is incorporated into historical narrative is more subjective, as Jenkins argues, and is open to diverse interpretations. Here, though, meaning might be construed as truth in the sense that it conveys wisdom, a more capacious notion that certainty.

Jenkins's presentism is symptomatic of the disquiet of our times about how past and future relate to the present, and of the indeterminacy of the approach to history that he advocates. He expects little of the past as a frame of reference for distinguishing past from present, for he is anxious to move on in dealing with our present historical predicament on its own terms. He offers an exhortation to write history anew, but with the understanding that its interpretations be perceived as provisional proposals for making sense of the human world amidst the fast-moving flux of present realities.

MICHEL FOUCAULT AS MEMORY ICON

Interest in the life and work of French scholar Michel Foucault (1926–1984) serves to illustrate this tension between representation and experience as modes of memory. Foucault was a philosopher who found his

---

[9] Nicholas Wade, *Before the Dawn; Recovering the Lost History of Our Ancestors* (New York: Penguin Press, 2006), 1–11.

way into history. He was greatly admired for his independence of mind, overturning conventional modes of scholarly discourse. He made his reputation through a series of historical studies about public commentary on madhouses, hospitals, prisons, and other forms of social management. Foucault was interested in the discourses that represented practices of these institutions rather than the practices themselves. The proposition that historians deal in representations rather than realities became the signature of his method. The representations that most interested him were those that disrupt rather than confirm what historians had long perceived to be the continuum of human experience. In his *Archaeology of Knowledge* (1973), he rejected the method of historians of ideas—tracing the development of ideas from their origins—and called instead for a genealogical reading of intellectual discourse backward from the present. The effect was to challenge the idea of intellectual continuity—construed as a heritage upon which the present builds—in favor of highlighting discontinuities in intellectual representation, and so dissolving long-standing notions about the relationship between past and present in cultural history. The patterns of the past, he maintained, are to be found not in its cultural traditions but rather in the way texts out of the past are imported into the discourse of the eternal present. History is the record of such cultural production.[10]

For Foucault, therefore, there can be no appeal to a master narrative. The patterns of historical interpretation do not correspond directly to the existential timeline of human experience, but rather to those of constructed representations of the past. We read the phenomena of the world as they are embodied in texts. In searching for connections within this web of intertextuality, historical interpretation is an ongoing project of construction and reconstruction, and the form a historian's narrative assumes reveals the give and take of relating textual references. Odd textual juxtapositions, moreover, may conjure up new histories, in which discordant perspectives encounter one another to generate new meanings.[11]

---

[10] Michel Foucault, *The Archaeology of Knowledge* (New York: Harper & Row, 1972), 135–48.

[11] Issuing from Foucault's line of inquiry was the "new historicism" movement, a venture whose leading spokesmen were literary critics rather than historians. The term is a misnomer, for this historiographical current was not the old historicism revisited, but rather repudiated. Catherine Gallagher and Stephen Greenblatt, leading proponents, called attention to the cultural negotiation involved in the interplay among textual references. Catherine Gallagher and Stephen Greenblatt, *Practicing the New Historicism* (Chicago: University of Chicago Press, 2000), 1–19.

Subjective experience is off-limits, consigned to the speculative province of the imagination. Foucault's work, and that of postmodern theorists whose spokesman among historians he turned out to be, introduced a strong, and for many a suspect, note of relativism into historical scholarship. Each age, Foucault proposed, reinvents the past in its textual narratives, dispelling the illusion of continuity and challenging each age to wrest from the past usable representations for explaining its present predicament.

Nearly everyone found Foucault's work immensely stimulating. He was lionized for his provocative ideas and his original approach to interpreting history, to such a degree that he became an intellectual celebrity for his times.[12] But among many historians he was suspect for his radical insistence that textual representation is impenetrable as a route to the subjectivity underpinning these characterizations. Foucault set sharp limits on the degree to which the subjective intentions of authors may be elicited from their texts. In a provocative essay, "What is an Author?" (1969), he pointed out that the tight identification of authors with the substance of their texts is a modern convention. Throughout much of history, authorship had often been anonymous, or at least was considered less important than the knowledge the text conveyed. Once composed, a text acquired autonomy of its own.[13] This observation correlates with Foucault's personal vexation over efforts to analyze the subjectivity behind his own authorship. He preferred the notion than an author stimulates a line of intellectual inquiry among scholars about where their shared research might lead. Foucault's argument about the limited access to the mindset of the author rendered the notion of the self an enigma, a topic to which he himself turned in his later writings under the guise of investigating its "techniques," that is, methods for its interrogation. He turned to the ancient Roman and Greek philosophers who sought consolation in the notion of the care of the self.[14]

Ironically, Foucault's late-life inquiry into techniques of self-care incited a fascination among scholars about how he understood his own

[12] Patrick Hutton, "The Foucault Phenomenon and Contemporary French Historiography," *Historical Reflections* 17 (1991), 77–102.

[13] Michel Foucault, "What is an Author?," in *Language, Counter-Memory, Practice; Selected Essays and Interviews by Michel Foucault*, ed. Donald Bouchard (Ithaca: Cornell University Press, 1977), 113–38.

[14] Michel Foucault, *The Care of the Self* (New York: Random House, 1988); idem, "Technologies of the Self," in *Technologies of the Self; A Seminar with Michel Foucault*, ed. Luther Martin et al. (Amherst: University of Massachusetts Press, 1988), 16–49.

subjectivity. No sooner had he died in 1984 than scholars began to investigate his life history. Several biographies appeared over the following decade, all seeking to ascertain hidden dimensions of his personality and his view of life. Who was the man behind the mask of his writings, a mask Foucault himself argued cannot be removed? The interest suggests that Foucault may have overreached in his theoretical claims about the autonomy of the text. The experience of life could not be so easily contained in its representation, and in Foucault's case especially so because he was such a singular personality. All of his biographers point to his brilliance and originality in his endless pursuit of new departures from what he had done before, differences taking precedence over similarities in the historical chain of endless textual reconfigurations. Foucault was dynamic and creative, moving among varied interests, always in search of innovative ways to present them. The notion of a "final Foucault" was the title of a collection of essays published shortly after his death, including what was his last interview.[15] But editions of his lectures and other unedited writings continued to be published over the following decades.[16] All of these varied writings were carefully inventoried. But the question raised by Foucault himself—what is an author—incited his biographers to transgress the boundary that he claimed should not be crossed. The question remained: who was Michel Foucault, and how is he to be remembered?[17] Foucault's rhetorical strategy of self-effacement incited among his readers a fascination with his hidden self, and it would stimulate even more discourse in the search for the secrets of his life. In the end, Foucault himself became a "memory figure" for intellectuals.

Historian Jerrold Seigel (b 1936) was among the first to pick up on the quest to fathom the subjective Foucault. He wrote "Avoiding the Subject: A Foucaultian Itinerary," an article about Foucault's interest in subjectivity in his early scholarship, residues of which continued to appear in his

---

[15] James Bernauer and David Rasmussen, eds., *The Final Foucault* (Cambridge, MA: The MIT Press, 1988).

[16] Michel Foucault, *Dits et écrits:1954–1988*, ed. Daniel Defert and François Ewald (Paris: Gallimard, 1994), 4 vols.

[17] The editors of the *Collective Memory Reader* (Oxford: Oxford University Press, 2011), 252, point out how little Foucault himself wrote directly about the topic of memory, for the topic ran against the line of argument that he was propounding. In one lone reference in an essay on Nietzsche, he introduces the term "counter-memory" to characterize history as it sets itself free from the notion that memory should be its matrix.

later writings.[18] Seigel points out that the young Foucault subscribed to an idea about the radical subjectivity of the human condition. As he was coming of age as a scholar during the 1950s, he was influenced by humanist existentialism, though these early writings appeared in minor journals and were few compared with those of his mature years that received so much attention in mainstream publications.[19] His publications of the 1960s, notably *Madness and Civilization* (1960) and *The Order of Things* (1966), signaled a new intellectual departure, revealing his intellectual metamorphosis. Seigel sees Foucault's article on the obscure literary critic Raymond Roussel as crucial, even though it holds only a minor place in his authorship. Roussel was interested in language, especially for the way it links phenomena otherwise unconnected.[20] It was thanks to him, Seigel contends, that Foucault began to reflect more deeply on the power of language. His interest in "discursive practices" became the key to his method thenceforth. In one of his essays, he put the contrast succinctly by reversing the adage of the Enlightenment: it is not knowledge that is the basis of power but rather power that shapes knowledge.

Seigel explains how Foucault developed his notion of the subjective self indirectly by focusing on the way it was hemmed in by institutional forces that sought to objectify human behavior by holding it to conformist standards. In the public sphere, the self was framed by discourse about its nature. Rarely did the subjective self break free of these descriptive categories. Yet for all of his writings about institutions regulating human behavior, Seigel suggests, the liberation of the self is what mattered most to Foucault. The affirmation of radical subjectivity in his pre-1954 writings never disappears from his later work, though he couches that notion in ways that hide his motive. As a homosexual, Foucault saw himself as an outsider, and he was sensitive to the way society at large intrudes into personal privacy, seeking to undermine the autonomy of the self in demands for conformity to its conventions. That is why in his later years he turned to the topic of strategies of self-care—the ways in which the self seeks to reaffirm its independence by defining its own rules of comportment.[21] Here he turned to the techniques of self-discipline devised

---

[18] Jerrold Seigel, "Avoiding the Subject: A Foucaultian Itinerary," *Journal of the History of Ideas* 51/2 (April 1990): 273–99.

[19] Ibid., 281–85.

[20] Ibid., 287–93.

[21] Ibid., 297.

by Stoic philosophers. Homosexual love as presented by Plato, Foucault observed, was not an objective condition of identity but rather a relationship that could foster a shared love of the pursuit of truth. Lovers engaged in techniques of self-care to that end. Foucault explained his views on the liberation of the subject through discursive practices of self-care in his late-life writings and, more explicitly and personally, in interviews given shortly before his death in 1984.[22] It was the discourse of self-care that constitutes the self, for the notion of a radically subjective self was a lure to draw readers into an interest that could never be satisfied. Inquiry into the nature of the self permits explication but not explanation.

Among Foucault's biographers, I contrast two to show what was at issue: that of David Macey with that of James Miller. Macey's biography provides close attention to detail. He worked conscientiously to canvass the events of Foucault's life as well as to provide synopses of his writings. Macey's story puts Foucault's endless peregrinations on display. Foucault was a wanderer and his travels to so many places inside and outside France mirror his restless intellectual journeys through so many topics of interest in cultural history. His life was an odyssey of many episodes. There is no coherent plot line to his life history, only an ongoing differentiation and refinement of his ideas as he moved from topic to topic and from place to place. He concludes that Foucault's story is one of an "unfinished life," so many were the avenues he might have pursued had he survived into old age.[23]

James Miller, by contrast, presents Foucault's life history as a journey whose coherence becomes visible as it moves toward a denouement in his last years, indeed during his last days.[24] As Foucault grew older, the personal passions that remained hidden and subdued within his histories of discourse about social institutions led to studies of discourse about sexuality and the self that lay close to still unstated personal preoccupations with his own identity. Miller's account, therefore, builds toward a crescendo of personal revelation and disclosure during his last days as he was dying of a neurological disorder that was in all probability AIDS. Miller makes much of Foucault's late-life journeys to California, where he taught

---

[22] Ibid., 298; See also Michel Foucault, "The Ethic of Care of the Self as a Practice of Freedom: an Interview," in Bernauer, ed., *The Final Foucault*, 1–20.

[23] David Macey, *The Lives of Michel Foucault; A Biography* (New York: Random House, 1995), 457–80.

[24] James Miller, *The Passion of Michel Foucault* (New York: Simon & Schuster, 1993).

for short stints as a visiting professor at Berkeley and where he became engaged in risky sexual behavior in the gay community of San Francisco. For his account of Foucault's last days, Miller relies heavily on revelations by Foucault's intimate friend and confidante Hervé Guibert. Guibert was a photographer and novelist several years Foucault's junior. Some seven years after Foucault's death, he published a short story and then a novel that contained thinly veiled accounts of Foucault's deathbed confession to him of his deepest secrets that had troubled him all his life. Since Guibert himself died of AIDS a few years after these publications, we shall never know what is fact and what is fiction in Guibert's story. Still, Guibert's claims are plausible, and Miller takes at face value what he had to say about Foucault's revelation of morbid memories of his childhood experience during the war years. All his life he had kept them hidden within himself, daring to approach them only through his writings about the techniques of representation.[25]

For Miller, Foucault's last days reveal the secrets of an identity he had all his life chosen to hide. The key to his life history lay in repressed trauma, and interpreting these memories provides insight into the topics he took up as author and the life that he chose to live. More than a method, Miller argues, Foucault's authorship was fashioned to serve his ideal that life itself should be lived for the experiences it offers, with all the risks that these entailed. For Miller, Foucault's late-life interest in the practices of the "care of the self" in antiquity was as close as he came to revealing his personal identity in his writings. He took Diogenes of Sinope, the ancient philosopher of Cynicism, as a model to emulate in pursuit of a lifestyle that tested the limits of life experience. Foucault tested those limits in his dangerous sexual experiences in California in the early days of the AIDS epidemic, and it led to an untimely death whose nature neither he nor most of his French associates were willing to admit. In view of these last days, Miller concludes, Foucault's life story takes on a tragic coherence. The story for some readers came to rival interest in his philosophy, or at least to put it in perspective. The fascination with his secret life suggests a reaction to his radical insistence on textual representation as an impenetrable barrier to subjective authorial intentions. [26]

My point is that the shift of scholarly interest from Foucault's theory of history to his personal life illustrates in an anecdotal way the unavoidable

---

[25] Ibid., 364–74.
[26] Ibid., 362–63.

tension between representation and experience as modes of appreciating the past. Moreover, it is in experience that the relationship between history and memory appears to draw closer together. What follows are some perspectives on the ambiguous relationship between the two. As Walter Benjamin argued, to revisit the experience of the past is to kindle its emotions vicariously, for these have an afterlife that may be evoked again and again. This idea provides a segue to our inquiry into memory's role in understanding history as experience as well as representation.

FRANK ANKERSMIT: REPRESENTATION VERSUS EXPERIENCE AS MODES OF HISTORICAL UNDERSTANDING

The tension between the writings of Michel Foucault about textual representation and the equally compelling interest in his inner life invites consideration of the work of Dutch philosopher of history Frank Ankersmit (b 1945), who has explored the relationship between historical representation and subjective historical experience. He wrote a book about each as modes of remembering that together contribute to his formulation of an aesthetics of historical understanding.[27]

Ankersmit greatly appreciates the value of a generation of theoretical work on historical representation, as agglomerated in the term "linguistic turn."[28] He recapitulates the contribution of these rich and thought-provoking studies, seeking to filter out what is useful from what is excessive. He lauds the initial forays of Hayden White into the linguistic foundations of historical narrative, before reaching back further into the writings of German philologist Erich Auerbach (1892–1957) on the concept of mimesis.[29] For the ancient Greeks, the term meant the imitation of nature. For Aristotle in his *Poetics*, the concept was full of ambiguity— reality as it is, or as it might be. Auerbach emphasizes the latter. The term in its most profound usage involves mimicry employed in a creative way to convey life experience. Mimesis, therefore, has an aesthetic foundation. Its value lies in the author's capacity to make of representation a work of art.

---

[27] Frank A. Ankersmit, *Historical Representation* (Stanford, CA: Stanford University Press, 2001); idem, *Sublime Historical Experience* (Stanford, CA: Stanford University Press, 2005).

[28] Ankersmit, *Historical Representation*, 29–74.

[29] Erich Auerbach *Mimesis; The Representation of Reality in Western Literature* (Princeton: Princeton University Press,1953), a seminal philological study of the historical elaboration of this concept, is minimalist as theoretical interpretation. His study has, nonetheless, been highly influential as an approach to the aesthetic fashioning of narrative.

Auerbach explained this notion as the figuration of reality, the imaginative shaping of its contours so that the representation conveys the past in its verisimilitude—a rendering that elicits from its audience an intuitive sense of what life was like back then.[30]

Ankersmit accepts the proposition that there can be no historical interpretation without its figuration in language. His desire, however, is to rehabilitate the idea of experience at a time when representation appears to have set limits on the possibilities of knowledge of the past. Ankersmit challenges the excesses of the language theorists, notably Richard Rorty and Jacques Derrida. He concedes that Hayden White's linguistic turn was one of the great historiographical breakthroughs of the twentieth century. But language, he argues, does not convey all that we may know about the past. The ability to recapture its feelings, to sense what life was like back then, remains an alluring magnet. Experience, Ankersmit contends, lies deeper than language. Experience is ever changing, but it provides our sense of continuity with the past. He believes that we can apprehend what life was like in the past not just as description, but also as imaginative empathy, with all of the attendant emotions that it evokes. It is not that we feel that experience again as it once was, but that we can comprehend what it must have felt like. He therefore asks the question: is it possible to find a way into that past that conveys the subjective experience of its historical actors in a way that is deeper than its linguistic representation? In its highest aspiration, Ankersmit believes, history can recapture intimations of the living past, a past that includes the range of human experience in the fullness of its visceral perceptions, imagination, and emotions. Put differently, his goal is to reconcile memory (the province of subjective experience) with history (that of textual representation).[31]

As a point of departure in this venture, Ankersmit returns to the work of his countryman Johan Huizinga (1872–1945), author of *The Waning of the Middle Ages* (1919), a study with remarkable staying power as a pioneering contribution to cultural history. Huizinga is famous for his capacity to rouse within his readers aesthetic images of the past. He aspires not only to explain the objective realities of those times, but also to convey some sense of subjective life experience—to imagine vicariously what life was like in those times. He aspired to conjure up the mood of the past, that is, to impart an understanding of the feelings of those who lived

---

[30] Ankersmit, *Historical Representation*, 197–217.
[31] Ankersmit, *Sublime Historical Experience*, 17–68.

in the Middle Ages. His descriptions lure readers into an imaginative reconstruction of life back then, with the understanding that some sense of lived experience may be communicated across the ages.[32]

To explain his method in terms of poetical tropes, Ankersmit suggests, Huizinga possessed the ability to move beyond metaphor in his figurations into the realm of synecdoche—the trope that conveys an overall impression of the past in its imagery. He sought to reach into visceral precognitive understanding of what it feels like to experience life. Ankersmit points to Huizinga's interest in synesthesia, the undifferentiated realm of experiencing the world that underlies the formation of language.[33] Synesthesia is the mode in which experience of the world is most deeply felt, and it is the ground of the perception of historical continuity. The idea of experience is all about the "presence" of the past. It is volatile and fleeting, most acutely conveyed by a sense of the sublime, aesthetic experience that is uplifting yet ineffable. The experience of the sublime implies immersion in the life world prior to the differentiation of subject and object, experience and language, a realm beyond the limits of representation. We cannot make the past come alive again. But we can reenact the past vicariously as aesthetic experience. Huizinga looked for "keys" to the past, by means akin to musical notes. Keys trigger moods, as in the experience of being transported by the harmonies of a great composer. The critical distance that we identify with historical interpretation is overcome by feelings of empathy for what was. He thought of such understanding as groundwork for its representation. Empathy with the historical actors of the past permits time travel.[34]

Ankersmit cites Walter Benjamin's notion of "aura" as indicative of the historiographical notion he wished to convey.[35] "Aura" is deeper than language; it conveys the immediacy of the original experience that we can come to appreciate vicariously. It is not that we feel the same thing but rather that we can understand what it was like to experience that past, and so to evoke its presence for our reflective consideration. Aura is a key to nostalgia. Our experience of the lifeworld changes, and in light of the new way in which we experience the world we become aware of our

[32] Ibid., 119–28, 133–39.
[33] Ibid., 128–33.
[34] Ibid., 225–27, 306–12.
[35] Ibid., 182–83.

departure from what we experienced back then. Such awareness enables us to discriminate between past and present. It makes it possible for us to believe that the past is stable, a reassuring referent for our lives in the present that is unstable in its uncertainties.

Is Ankersmit's project to join memory's experience with history's representations viable as a theory of history? A great deal of effort, particularly in recent decades of the memory phenomenon, has been devoted to understanding related practice based on such a conception. These include a range of practices—from historical novels, to theater, to tourist pilgrimages, to historical reenactments. Most of these practices have been around for a long time, with little critical attention to their relationship to history as practiced by professional historians. The media revolution has sensitized the present age to interrelationships among them as approaches to the past. Here history draws closer to the arts of memory. In a philosophical way, Ankersmit has opened the door to what has recently been labeled "performative history."

VARIATIONS ON THE QUEST TO RELIVE THE PAST VICARIOUSLY

The lure of vicariously reliving the past is an old if impossible dream. Nineteenth-century historians had little compunction about enhancing their analysis of evidence with the imaginative reach of their prose, and professional historians today are well aware that what once passed for history in these sometimes florid writings would now be labeled imaginative fiction. Even today, the desire to recapture the past as a living experience persists in many domains, revealing a divide between professional and amateur historians.[36] History buffs continue to be taken with the evocation of an imagined past. Reviewers in popular newspapers and magazines of commentary still praise historians who write about topics of perennial historical interest—for example, the lives of the American "founding fathers"—in a way that makes them "come alive again." The public longing to reexperience the past finds expression in the popular cult of historical reenactment of signal events, usually military battles of the

---

[36] Among professional historians, Harvard art historian Simon Schama is a master at drawing history as representation as closely as possible to history as experience, as in his *Landscape and Memory* (New York: Knopf, 1995).

American Revolution, the Civil War, or World War II.[37] Practitioners of historic preservation are likewise faced with the need to attract the public by drawing them into a reconstructed milieu in a way that conveys the illusion of time traveling into the past.[38] Today's mass tourism industry is based on the proposition that the return to physical places of memory quickens a feeling for what once transpired there. Good teaching in the public schools, moreover, is often equated with a teacher's capacity to engage students in an emotional involvement with their subject matter, for example, by showing up in period costume. Students of theater remark on the power of historical drama to enliven the historical imagination.[39] The appeal of the image making of television programming, notably on the History Channel (its now rare depiction of historical subject matter notwithstanding), originated out of just this need.

It is understandable that professional historians have looked with skepticism upon history as imagined reenactment. They contend that the quest to relive the past is a misguided appreciation of what history can tell us about it. Such techniques for promoting vicarious identification with the past are rather arts of memory, stimulating memory's flights of imagination, not history's grounded empirical analysis. The quest to reexperience the past is fraught with temptation to stray from hard evidence into soft fantasy.[40] Still, memory studies make manifest that poets, novelists, and artists—as well as historians—have something profound to say about the appreciation of the past, and that history as a discipline operates within a field of creative ways to extract its varied meanings. Some historians have begun to explore these connections.[41]

[37] For the mindset of historical reenactors, Jenny Thompson, *War Games; Inside the World of 20th-Century War Reenactors* (Washington, DC: Smithsonian Books, 2010).

[38] Diane Bartel, *Historic Preservation; Collective Memory and Historical Identity* (New Brunswick, NJ: Rutgers University Press, 1996).

[39] The topic is explored in both theory and practice by Roger Bechtel, *Past Performance; American Theatre and the Historical Imagination* (Lewisburg, PA: Bucknell University Press, 2007).

[40] Schama has his critics. In a review of Schama's *Dead Certainties* (New York: Random House, 1991), historian Gordon Wood chides him for straying too close to fiction. "Novel History," *New York Review of Books* 38/12 (27 June 1991).

[41] See the collection of essays on the "performative turn" in Karin Tilmans, Frank van Vree, and Jay Winter, eds., *Performing the Past; Memory, History, and Identity in Modern Europe* (Amsterdam, Netherlands: Amsterdam University Press, 2010).

ANN RIGNEY ON THE BLURRED BOUNDARY BETWEEN HISTORY AND HISTORICAL FICTION

The historiographical opening of students of performative history toward subjective experience explains the renewal of interest in historical fiction vis-à-vis history. Irish literary historian Ann Rigney (b 1957) explores that relationship in her ongoing work on the historical novel. She explores the way fiction shares many of the traits of history, yet as fiction can be more memorable in its ability to kindle the historical imagination. She addresses the way the line between fact and fiction has blurred in our age of media. Working on the boundary between history and historical fiction, Rigney analyzes differences in their appeal. History promises objectivity. Yet fiction may convey the authenticity of verisimilitude and in the end may make the past more memorable. She explains that the historical novel was in the forefront of new mnemonic practices in the early nineteenth century.[42] Over the course of that century, history as a discipline would distance itself from its literary heritage as it aspired to become a science. In our times, however, the line between them is not so sharply drawn. The issue at stake is not only which narrative is more accurate, but which one is more memorable.

If recourse to memory proceeds from the reference point of the present, Rigney asks, what makes one representation more memorable than another? Two factors, she suggests, are involved: the aesthetic power of the narrative and its cultural longevity. The irony, she notes, is that the fictive account may have a more enduring appeal than that which is closer to the factual record. The novel, for example, may have more staying power because it is a fixed target for scholarly criticism, whereas historical problems are constantly subjected to revisionist interpretation, redirecting attention to different evidentiary sources.

As a centerpiece illustrating her interpretation of the relationship between history and historical fiction, Rigney explores what might be characterized as the "afterlife" of Walter Scott's historical novel *Ivanhoe* (1820).[43] The novel was immensely popular not only in Great Britain but abroad as well, and its appeal endured well into the nineteenth century.

---

[42] Rigney, "The Dynamics of Remembrance: Texts Between Monumentality and Morphing," in *Cultural Memory Studies; An International and Interdisciplinary Handbook*, ed. Astrid Erll and Ansgar Nünning and Erll (Berlin: Walter de Gruyter, 2008), 345–53.

[43] Ann Rigney, *Imperfect Histories; The Elusive Past and the Legacy of Romantic Historicism* (Ithaca: Cornell University Press, 2001).

She shows how Scott's *Ivanhoe*, a story about the Middle Ages that contributed to Scottish nationalism by pitting native Saxons against invading Normans, was redeployed in nineteenth-century American Southern culture to romanticize its ideal of chivalry among the plantation aristocracy. Apologists for that culture fought valiantly in defense of a lost cause. She notes how the "Southern tournaments," so popular in the antebellum era, were precursors of modern day historical reenactments. As such they played a role in establishing cultural identity in the South. The novel became a touchstone for remembering the Southern heritage in an idealized way, and so gave it a cultural longevity it would not otherwise have known. Rigney investigates this legacy with a view to understanding the way the novel was read, and remembered imaginatively in a variety of cultural settings. The appeal of the historical novel, she argues, lies in its capacity to evoke the historical past as a living presence for its readers. The power of such reception is intimately tied to the novel's capacity for reenchanting the past and so stimulating the imagination. It dramatizes the past as if it were a theater of the memory of those times. So received, the novel lives on in cultural memory as an idealized representation of a time in history. Rigney plots the stages of the remembrance of *Ivanhoe* as a poetical logic. It follows a trajectory that slopes away from familiarity. What began as a concrete fixed point of reference: a story about the Middle Ages, becomes telescoped into the totality in its title, before being reduced to an icon deprived of its literary meaning, finally passing into obsolescence. The pattern that she explains is reminiscent of Giambattista Vico's formulation of the poetic logic of the figures of speech, following refigurations in the pattern of remembrance as opposed to sequences in the framing of the chronology of the events of history.[44]

In this argument, Rigney follows the model of Jan Assmann's mnemohistory. Novels, like all works of art, may be assessed for their place on a graded scale of their memorability for posterity. The more memorable a work in the judgment of following generations, the more likely it is to acquire canonical status. Such an argument challenges historicism's criterion for evaluating the meaning of a work of art. Historicist theory contends that such meaning must first be appreciated in the context of its particular time and place in history. Rigney, by contrast, wants to show how a novel may take on new meanings as it is appropriated in other

---

[44] Rigney, "The Many Afterlives of Ivanhoe" in *Performing the Past*, ed. Karin Tilmans, 207–34.

contexts with the passage of time. Her idea of historical remembrance cor-relates closely with the notion of a mnemonic afterlife. Here the historical novel, far from being appreciated in the close reading of its narrative for what it may disclose about the specific meaning of the time and place of its setting, may find larger meaning as it is remembered in condensed form as a "memory figure" for audiences in other times and places. From this perspective, the historical novel acquires a performative role to serve some present-minded purpose for later generations. In this way, she argues, a literary work becomes an "active agent" of cultural remembrance. It may have enduring importance not so much for what it says about its own times as for what the present-day culture into which it is imported finds worthy of remembrance. A memorable historical novel, therefore, may have many lives.

Rigney argues that one can trace a chain of memory in such appro-priations, a sequence apart from that of the historical chronology of the novel's actual distribution. Different groups may find different incentives for imagining their affiliation with personalities or groups portrayed in the novel. Here, she explains, representation of the past typically becomes highly selective. The past is telescoped to put the accent on features of the past that the novel portrays effectively. Historical remembrance, therefore, tends to rely on figurative representation. It showcases only a few memory figures out of the myriad possibilities of those times. Plot and character become concentrated in iconic representation, reinforcing their worthi-ness for remembrance. Such icons serve as "touchstones," and are easily refashioned to conform to the way later-day readers wish to imagine the relationship between that past and their own present.[45]

François Hartog: Presentism as Today's Model of Historical Time

Historian François Hartog (b 1946) provides a historiographical context for assessing postmodern approaches to history, with their presentist per-spective and emphasis upon historical experience. In his study, *Regimes of Historicity* (2015), he argues that it is our newfound perspective on historical time that explains the historians' openness toward the dynamics

[45] Ibid., 224–29.

of collective memory.[46] Ours is a present-minded age, he explains. But it is one that has emerged only in the last third of the twentieth century, and accounts for our sense that we have left the modern age behind.

Presentism may currently be in vogue, he contends, showcasing collective memory and diminishing the place of history among the arts of memory. But there are limits to memory's claim upon history, if their relationship is considered over long periods of time. But presentism need not be viewed as the culmination of advances in our understanding of historical time that put earlier approaches to rest. Hartog attributes presentism to a crisis in our conception of historical time, a response to the waning appeal of its modern formulation as a linear directional narrative. The future as conceived in the historical scholarship of the modern era, with its upbeat anticipation of progress, no longer speaks to present conditions. The sense of continuity between past traditions and present projects has been disrupted by new historical realities; the future is beclouded in uncertainty about the human prospect. Presentism, Hartog contends, is another way of saying that our conception of historical time has come to reside in the immediacy of a short-term perspective on it.

Two historians especially stimulated Hartog's thinking about this topic: Pierre Nora on the crisis of French historiography with its recourse to places of memory, and Reinhart Koselleck in his formulation of the making of the modern conception of historical time in the late eighteenth century.[47] Hartog weaves their insights into a larger context. He is himself a student of historical writing in the ancient Greco-Roman world as well as nineteenth-century historiography, and so has been able to stake out changing conceptions of historical time through the ages.[48] He labels these "regimes of historicity," and he identifies three: ancient, modern, and contemporary.[49]

The ancient regime, Hartog explains, was conceived as a *historia magistra vitae,* a "magisterial history of edifying lives." It privileged the past

---

[46] Hartog, *Regimes of Historicity; Presentism and Experiences of Time* (New York: Columbia University Press, 2015), trans. of the French ed. of 2003.

[47] François Hartog, "Temps et histoire; 'Comment écrire l'histoire de France?'" *Annales HSS* 50/6 (1995): 1219–36.

[48] Hartog on historiography of antiquity François Hartog, *Mémoire d'Ulysse* (Paris: Gallimard, 1996); *Le Miroir d'Hérodote; Essais sur la representation de l'autre* (Paris: Gallimard, 1991); *Le XIXe Siècle et l'histoire: Le Cas Fustel de Coulanges* (Presses Universitaires de France, 1988).

[49] Hartog, *Regimes of Historicity*, xvii–xviii, 15–18.

as the defining moment of historical time, remembered as a golden age to which the present can never measure up. Until the Renaissance and even beyond, historians looked to past precedent. Legendary heroes out of the past were revered as models to emulate. History was taught by example and analogy.[50]

The future-oriented modern regime of historicity, Hartog argues, dates from the late eighteenth century, more specifically the era of the French Revolution. It looked for the origins of historical trends and emphasized continuities between past and present. As a historical perspective on time it embodied the great expectations of the future envisioned by the philosophers of the Enlightenment. The Revolution was seen as such a point of origin for this reorientation, a harbinger of better times to come. Nineteenth-century historians, therefore, took the future as their primary referent of historical time and the rise of the nation-state as the main character of their story. So conceived, historical time was understood to unfold as a linear storyline, its origins significant for the direction it set, its key events as turning points, denoting stages along the way of its trajectory. Modern historiography was written as a grand narrative of humankind's advance toward a presumed destiny. Such a view of historical time prevailed for the first two-thirds of the twentieth century.[51]

Our presentist conception of historical time, Hartog continues, surfaced toward the end of the 1960s. It was precipitated by a loss of a sense of continuity between past and present. Hartog notes a number of features contributing to this disruption, notably trends toward globalization and the waning fortunes of the welfare state. Couple these with a sense of the acceleration of time promoted by the media revolution and the horizons of expectation for the future collapsed into the immediacy of present concerns. All the while, the timeline of the grand narrative of the rise of the nation-state was breaking up. The past as carried forward in modern traditions was losing the force of its once defining meaning. The notion of destiny gave way to one of uncertainty about what the future may hold. Its prospects no longer appeared auspicious, and were recast as portents of looming dangers. The immediacy of the present as the determining moment of historical interpretation came to the fore as the principal referent for historical time. The distinguishing trait of today's historians of the

[50] Ibid., 72–77.
[51] Ibid., 131–41.

present age is their confidence that they can acquire a critical perspective on a present age in which they themselves are immersed.[52]

The entry of these regimes into historical thinking, Hartog contends, took place in times of disruption that produced a sense of breaking points in the timeline of human experience. They altered expectations about the meanings to be sought through historical interpretation. He labels them gaps rather than transitions to avoid connotations of an underlying continuity in history's narrative. To explain the ideas implicit in these regimes, he points to historians who reflected on crises in thinking about historical time at such moments in the past: Augustine for the ancient regime; René de Chateaubriand for the modern one.[53] Hartog, therefore, sees himself in an analogous position, articulating the changing meaning humankind is investing in the crisis of our time in history. He is aware of the affinities between his own project and that of the classical philosophers of history, from Nicholas de Condorcet to Arnold Toynbee. But in addressing historiographical conceptions of time rather than patterns in the events of history, he sidesteps the pitfalls into which those philosophers fell in their efforts to encompass the past in a deterministic overview of stages along the way. Philosophies of history are wildly speculative about history's narrative, whereas regimes of thinking about historical time are grounded in the texts of the historians themselves and reflect their subjective perspectives. His thesis, therefore, is more modest, but at the same time more useful for understanding the current crisis of historiography, in which history appears to yield its once preeminent place among the arts of memory. His work reflects the degree to which historiography—conceived as the history of historical writing—has come to the fore in the interests of today's historians.[54]

In my reading of Hartog's interpretation of a presentist regime of historiography, I note the following features, some stated, others implied:

---

[52] Ibid, 107–14.

[53] Ibid., 65–99.

[54] On presentism as a conception of historical time, see also Chris Lorenz, "Unstuck in Time. Or, the Sudden Presence of the Past," in *Performing the Past*, ed. Tilmans, 67–102. Lorenz includes commentary on Hartog's thesis.

Genealogical Descent into the Past

One reads the past retrospectively from the present in the manner of a genealogy. Such a reading highlights discontinuity between past and present, in opposition to the continuity valued by modern historiography.

The Spatialization of Historical Interpretation

One orders past and present synchronically rather than diachronically. The points of departure for interpreting the past are topics rather than origins. Narrative proceeds from these places of memory, of which there are many, obviating the notion of a metanarrative that integrates them all. Historical inquiry is a pluralistic enterprise that easily accommodates incongruent interpretations. The breakdown of unified traditions of historical writing has permitted the rise of *égo-histoire*, a historiographical phenomenon of the 1980s in which historians wrote confessional memoirs of their initiation into the profession as a manifestation of their personal autonomy and independence as interpreters of the past.

The Acceleration of Time

The revolution in the technologies of communication has multiplied exponentially the publicizing of events, and so has reinforced presentist perceptions of the ever more discrete segmentation of time. Presentism also puts its accent upon change wrought by science, technology, and cultural innovation, as opposed to the inertial power of the past embodied in habits of mind and traditions based on custom. The effect is to create the impression that time is speeding up.

Past and Future Enveloped in the Expanding Present as a Regime of Historicity

The meaning of the present as a moment in historical time has come to be characterized by its indeterminacy. The diminished faith of the present age in the past's precedents and the future's promise has dissolved the sense of continuity essential to the idea of a single timeline of history. History can no longer be conceived as a grand narrative, but as a host of discrete, and not necessarily congruent, narratives that historians choose to address in their random travels back in time. Such a perspective breaks up the sense of continuity that informed the understanding of historical time in the modern regime of historicity. It is not as if past and future no longer matter in our presentist regime of historicity. In some ways, they matter more than ever before. If the past is no longer perceived to exercise an inertial power, its uses in the present are understood to be more open-ended. The horizons of the present age are wider, even if expectations of the future are no longer clear. The past is therefore revisited as a resource for re-visioning the future in terms of its infinite possibilities. As the future becomes less predictable, so too does the past. It becomes strange, a "foreign country" to be entered in more tentative ways.

Caution vis-à-vis the Future

The future, in turn, has waxed large in the apprehension it arouses. Hartog invokes Immanuel Kant's principle of the categorical imperative—act as if you were acting in behalf of all humankind. Kant's future from the vantage point of the early nineteenth century was one of anticipation. One might take risks in the name of bettering the human condition. While the principle remains the same, Hartog notes, the nature of the action advised in the presentist regime of the early twenty-first century is redirected toward caution. In a world as complex and dangerous as ours has become, one must act with careful planning and discretion not only to improve the human condition but also to repair its failings. With no specific notion of a destiny in the offing, the burden of ethical decision falls more heavily than ever before upon the leaders of the present age to make responsible choices.

Historic Preservation and the Heuristic Effect of Memory

Memory has in the presentist regime of historicity become a preoccupation, even an obsession. But memory, in the modern regime conceived as tradition (a sustaining past), has now been reconceived as heritage (a useful resource). Memory today, Hartog argues, has asserted its claims upon past and future in the movement for historic preservation of the built environment. The movement to preserve its treasures was institutionalized in heritage sites during the 1970s, and in university programs during the 1980s. The movement not only burgeoned but diversified. For if specific events no longer provide preconditions for the emergence of the present, as they did in the commemorative rituals of the modern regime, then any and all mementos of the past are potentially worthy of remembrance. The preservationists' duty is to preserve as much of the built environment as might possibly be judged memorable, even in our minimalist expectations. Here Hartog's remarks about the preservationists expanding conception of the scope of preservation resonates with Aleida Assmann's remarks about the expanding archive of conservationists of the documentary record. The preservation movement had grown exponentially by the end of the twentieth century, its interests ubiquitous yet diverse in their localization.

The preservation movement has stimulated rethinking projects of commemoration. The past is not to be mourned, nor should its ruins be thought of as place marker on the linear timeline of modernity's history, but rather as potential sources of inspiration that may come alive again in their restoration. As heritage the past bursts into the present. In this setting, the past is experienced vicariously in ways that stimulate the imagination about what the past was like, taking into account not only its ways of thinking rationally but also the feel of its emotions.

In one of his most original insights, Hartog links cultural with biological preservation.[55] The rise of the environmental movement to protect nature, he observes, is coeval with that of the movement to protect the built environment. For the latter, the stakes are larger for our present concerns. Unless we act now, scientists caution, the consequences for the future of the biosphere are dire if not catastrophic. Such warnings deepen the dilemma of presentism. In an age in which short-term economic interests override long-term environmental planning, how does one persuade the public to sacrifice immediate satisfactions for long-range goals tending toward a distant future whose nature is indeterminate and difficult to envision. As the economist and futurologist Robert Heilbroner once tellingly quipped about the all too human presentist mindset: "what has posterity ever done for me?"[56] Hartog's point is that we are caught up in a regime of historicity that favors immediacy. Presentism in practice offers little concrete incentive to invest in a future that we cannot yet envision. Nor can it countenance what history's future role may be.

---

[55] Hartog, *Regimes of Historicity*, 149–54, 186–91, 193–204.
[56] Robert L. Heilbroner, An *Inquiry into the Human Prospect, Looked at Again for the 1990s* (New York: Norton, 1991), 183.

CHAPTER 8

The Memory Phenomenon as a Time in Historical Scholarship

The Royal Roads of History's Inquiry into Collective Memory: A Recapitulation

Scholars argue that there have been three waves to the memory phenomenon: The first in the late nineteenth century was scientific and focused on the individual. Its realm was psychology. The second was in the late twentieth century and lay closer to the arts of memory. It focused on the social. Its domain was historiography. The third is now upon us, having emerged in the early twenty-first century, and focuses on the cultural implications of advances in media technology. Its realm is communication science. Memory studies today ride this wave.

In my study, I have concentrated on the second wave, which was triggered by a crisis in historiography. But I have done so with an eye on the third, for it is in this scholarly milieu that the close interchange between memory and history has drawn them into a churning mix with the breaking of the wave. In my search for a context, I have identified three royal roads of scholarship: commemorative practices; the cultural implications of transitions in the technologies of communication; and the disabling effects of trauma, with particular attention to the memory of the Holocaust. Emerging during the 1980s and converging by the 2000s, these approaches to memory studies have been reordered as we approach our present circumstances. In 1984, the problem of memory was addressed under the aegis of historical inquiry. Today it is memory

© The Editor(s) (if applicable) and The Author(s) 2016
P.H. Hutton, *The Memory Phenomenon in Contemporary Historical Writing*, DOI 10.1057/978-1-137-49466-5_8

studies that accommodate history as one among the arts of memory. Here is a brief review of the trajectory that I have followed:

Commemorations

The first among these was the study of commemoration. This was the obvious point of departure. Commemorative practices were of immediate interest to the crisis of historiography. The old places of memory as prompts for the study of history—class conflict with the ascendancy of the bourgeoisie, the role of the nation-state under the banner of and the idea of progress—had lost their appeal. The sense of their irrelevance to the project of historical writing in the late twentieth century was symptomatic of the breakdown of the metanarratives of nineteenth- and early twentieth-century historiography. The historians' perspective on this crisis of historiography found expression in such provocative notions as the "end of history" or the coming of a "postmodern age." Neither concept was adequate to explain the crisis comprehensively. But their use in historiographical discourse signaled the loss of the temporal framework that the master narratives of modern history had once provided as foundations of collective identity. National history was preeminent among them, inspired by a past still cherished, but no longer with the same naiveté.

Historians had become reflective about the nature of commemoration, and they recognized that commemorations have a history worthy of study. Eulogy was replaced by autopsy, as initiated in Hobsbawm and Ranger's much appreciated study of the "invented" tradition. Scholarly attention was directed to the politics out of which commemorations were inaugurated, modified over time, and eventually contested as their influence waned. Ironically, such challenges contributed to the longevity of commemorative traditions. Herein history claimed mastery over memory. Today such studies, with their attention to fixed places of memory—monuments, museums, places of pilgrimage—continue to inspire scholarship, though they are no longer at the forefront of work in memory studies.

Holocaust Remembrance

Studies of the remembrance of the Holocaust of European Jews have followed a second major route into study of the relationship between memory and history. During the 1980s, they dominated German and American historiography of the memory phenomenon. Work in this field had a decidedly historiographical bent because it revealed the degree to which issues of unrequited memory retard the task of historical interpretation. A solid foundation of evidence concerning this shameful episode in modern European history had been well-established. But the question of how the Holocaust should be remembered remained an open and divisive issue. Scholars recognized that there were gaps in the record because the testimony of its victims was so difficult to evoke in light of the trauma they had experienced. Here memory offered resistance to history's demand that it give up its secrets, making it difficult to move toward a comprehensive interpretation of the meaning of this episode in modern history. So the psychoanalytic task of recovering repressed memories continued, while sharp historiographical controversies over the meaning of the Holocaust left interpretation interminably unresolved.

The history of the Holocaust, especially for the ways of its historical remembrance, therefore, remains a vibrant field of scholarly interest, especially as it has become institutionalized in university curricula. Though the Holocaust may be viewed as exceptional as a crime against humanity, it has come to serve as a model for the investigation of the disabling effects of trauma upon historical interpretation. In recent years, it has stimulated interest in the resistant effects of trauma upon the interpretation of genocides of the late twentieth century, in Bosnia or Rwanda, and in earlier historical episodes, such as the Terror during the French Revolution.[1] As an obstacle to historicization to be decoded, the problem of trauma receives widening scholarly attention.

Media Studies

Media studies, our third avenue of inquiry into the memory phenomenon, originally focused on transitions among oral tradition, manuscript and print literacy, and initially provided a less traveled road into the history of memory. Certainly it was the media revolution of the mid to late twentieth century that sparked this interest, though scholars turned first to analogous transitions in the technologies of communication deep in the

---

[1] See the varied case studies analyzing traumatic memory in relation to historical in the anthology edited by Michael Roth and Charles Salas, *Disturbing Remains: Memory, History, and Crisis in the Twentieth Century* (Los Angeles: Getty Research Institute, 2001); and in the special issue "Trauma and Other Historians," edited by Yoav Di-Capua, in *Historical Reflections* 41/3 (2015).

past. From the 1960s to the 1990s, media studies directed attention to the cultural effects of orality versus literacy. Such scholarship was the province of classicists, folklorists, biblical scholars, anthropologists, and subsequently historians in their inquiries into the threshold between the two. They devised new methods for ferreting out oral residues within written texts, thus permitting historians to reach back into the culture of primary orality. Studies of print literacy by historians followed, with particular attention to the Enlightenment's "republic of letters." Print technology vastly expanded the production of texts of all kinds available to the public and democratized reading habits in the process. Here too, history initially demonstrated its mastery over memory. One could document the historical transitions in technologies of communication from primary orality to print literacy, and explain the changing cultural practices that followed from them. In scanning these changes from antiquity into the nineteenth century, orality/literacy gave rise to a master narrative of its own. The advent of new technologies signaled points of demarcation in the long history of cultural communication and so provided a different kind of model for historical periodization.

Historians have been slow to take up the most recent media revolution, the workings of memory in the late twentieth-century electronic revolution, and so by default relinquished the task to communication scientists who did so with vigor by the first decade of the twenty-first century. They, together with scholars hailing mostly from programs in literature, would make their case for memory's autonomy in the age of digital communication. By the turn of the twenty-first century, the topic of orality/literacy had faded from the scholarly spotlight while digital age memory came to be studied with a rush of intensity, as if it were the only medium of communication that mattered. It is this pathway that has expanded into the superhighway of scholarship on cultural memory in our times. Historiography was not so much subordinated as marginalized. The label for work on the memory phenomenon dubbed "history and memory," commonplace in earlier decades, yielded place to that of "memory studies," signaling the presentist perspective this interdisciplinary venture affirmed. Unbound from history, media studies celebrated digital memory's dynamic, transforming powers. In the transition, history continues to find a modest place on the margins of this new approach to memory scholarship, now one among the many arts of memory.

Jeffrey Olick, Vered Vinitzky-Seroussi, and Daniel Levy Review a Century of Memory Work Across the Curriculum

Early in the twenty-first century, a team of sociologists, Jeffrey Olick, Vered Vinitzky-Seroussi, and Daniel Levy, assessed the significance of the memory phenomenon in an expansive review of its accomplishments over the preceding century, published as *The Collective Memory Reader* (2011).[2] Their study serves as a landmark in this rapidly evolving field. Their conceptualization places scholarship in history within a widening context of research across the social sciences and the humanities. Rather than drawing toward a close, they suggest, interest in the dynamics of collective memory has taken on new energy and a more far-reaching allure. Their introductory essay to this anthology provides a comprehensive assessment of accomplishments in this field among scholars across the curriculum at the outset of the twenty-first century.[3] Far from fading away, they anticipate that memory studies in the digital age will only expand and take new directions. As a bright prognosis of the future of the interest in collective memory, their interpretation invites comparison with the more elegiac one presented by Nora in 1984. In light of their forecast, my question is, how will the earlier historiographical interest in the relationship between memory and history fare within today's interdisciplinary realm of memory studies?

So much had changed over the course of the golden age of memory studies. Nora had conceived of his project as an experimental venture. Olick/Vinitzky/Levy's project was a summing up of all the scholarship published in between, not just in history but across the curriculum. The field of memory in history had been transformed into memory studies. The *Collective Memory Reader* marked that passage into a wider scholarly world. Theirs was a much broader, more diverse investigation, one more confident about its purpose in comparison with the tentative beginnings of Nora's entry into the field.

There are technical ways in which the two projects are similar. Both were conceived as collegial ventures. Nora invited some 125 authors to join him in his project; Olick/Vinitzky/Levy some 100. Both were inquiries into the

---

[2] Jeffrey K. Olick, Vered Vinitzky, Daniel Levy, eds., *The Collective Memory Reader* (Oxford: Oxford University Press, 2011).

[3] Ibid., 3–62.

nature of collective memory, setting individual memory aside. Both looked to Maurice Halbwachs as a figure who had established the conceptual foundations of the field. Both might be regarded as manifestos about the state of scholarship in an emerging field of scholarship. Together they provide bookends to the mnemonic turn in history in the late twentieth century.

Yet in format, purpose, and thesis, these are quite different projects. Nora's anthology was a bold overture, original in concept and setting a new course for scholarship in French history, but with obvious implications for other national traditions of historiography. Its influence would spread in the manner of an expanding circle. Olick/Vinitzky/Levy's anthology contended that memory studies was not a venture nearing completion of its assignment but rather one redirecting intellectual inquiry from historiography toward the transmission of culture, taking the field in directions that transcended national boundaries and that surpassed Nora's conception of what the study of memory among the social sciences might be.

For Nora, inquiry into collective memory had been a response to a particular crisis in French historiography that arose over the course of the 1970s, precipitated by the breakdown of the French Revolutionary tradition as the matrix out of which the narrative of modern French history had been written. It represented at best a decade of work carried out almost exclusively by French historians of a single generation. This was a closely bound network of scholars, most of whom knew one another personally.

For Olick/Vinitzky/Levy, the interest in the history of memory was reconceived as a broader movement in scholarship, springing not from a mere decade of work in historiography but rather from that of a century across the spectrum of scholarly disciplines. In contrast to the interpretation of the memory boom as time-specific—for Nora the decade 1970 to 1980—Olick/Vinitzky/Levy date the origins of the movement to the end of the nineteenth century, presenting it as a movement that coalesced gradually over the course of the twentieth century, gathering momentum all the while. While Nora's inquiry had focused exclusively upon France, Olick/Vinitzky/Levy's project was international in scope, interdisciplinary in method. Initially, few of the leading scholars in the field worked together. They hailed from different places around the globe, and worked within different intellectual traditions. By the early twenty-first century, though, scholars in digitalage memory studies had established a network of communication in conferences and journals in Europe, with satellites in the USA and the Pacific Rim. In their project, the place of memory within historical studies gave way to the place of history within memory studies.

The issue was no longer the rise and fall of a field of inquiry, but rather an intellectual inquiry reconceived as it gathered strength and expanded its domain, integrating autonomous strands of research along the way to intermingle in an international scholarly forum. This coalescence of scholarship created synergy that for its editors presaged the development of a general theory of the role of collective memory in cultural communication.

Nora had guided the development of his project closely. He defined the rubrics of classification, and prefaced each with his own overview. He fitted the work of individual scholars into his scheme. For Olick/Vinitzky/Levy, by contrast, contributors speak for themselves. Among their varied approaches, the editors searched far and wide for connections, mixing studies from across the curriculum in broadly conceived categories: classics, identity, power, modes of transmission, and justice. They introduced their anthology with an overview that acknowledged the many strands of the memory phenomenon, to be understood as a network in the process of elaboration. Prefacing each entry with only brief orientation, they allowed contributors to present themselves in well-chosen excerpts from their writings that conveyed the gist of their point of view. Their format was representational rather than hierarchical, a memory plane as opposed to a memory palace.

For Nora and his colleagues, collective memory concerned a world fixed in the past, lost but now to be recovered: the deep roots of the French republic, nation, and culture. For Olick/Vinitzky/Levy, collective memory played out in a transnational milieu in the process of becoming, dynamic in the media resources it employed to transform the culture of the contemporary age.

For Nora, the memory phenomenon emerged out of a breakup of a linear model of history founded on the moral imperatives of the French Revolution. By the 1970s, it had become a fading tradition in the face of new historical realities. Expectant in the development of the pattern of history that this model of history anticipated, its breakup produced a sense of disorientation and a need to put it back together by looking into its sources in collective memory as a prelude to its reconstruction. France's era of commemoration, Nora argued, was a time of *attente* in which the old political metanarrative of French history was set aside in favor of reexamining France's cultural heritage. For Olick/Vinitzky/Levy, it was not memory as it had faded but memory as it is regenerated and recycled that captures the attention of scholars in our contemporary age. They point to memory's power to integrate the old into the new in keeping with its present-minded perspective. The study of memory today is not so much representative of a time in history as an investigation of the variety of ways in which the idea of temporality may be conceived. Memory studies from their perspective are more about coming to terms with a crisis of culture rather than of history, as made manifest in such themes as the politics of regret attending memory of the Holocaust; the commodification of nostalgia that enlisted sentimentality in the service of commerce; the cultural implications of the disintegration of the material archive as data was transported into the malleable realm of cyberspace. A summing up of a century of work, Olick/Vinitzky/Levy's *Collective Memory Reader* was a prelude to a decade of fast-moving developments in the understanding of collective memory in the digital age.

HISTORIANS ASSESS THE MEMORY PHENOMENON

Whither history in light of this evolution of the memory phenomenon? It is an open question. Historians have and continue to learn much from the study of memory, in which ideas about cultural memory continue to diversify.[4] But let me present synopses of the responses of five historians who have reflected on the memory phenomenon along the way, each contextualizing the answer differently: Gavriel Rosenfeld on the waning of the memory boom; Arlette Farge on working in the archives in the old way; Robert Darnton on saving the book in the age of the Internet; Yosef Yerushalmi on the tragic divide between history and historical remembrance; and Paul Ricoeur on the opposing vocations of memory and history.

Gavriel Rosenfeld: Looming Crash or Soft Landing?

Historian Gavriel Rosenfeld responds intuitively to the fact that the historians' inquiry into collective memory was becoming a crowded historiographical field and wondered whether after 30 years the topic was losing its appeal. In 2009, he published an article about the memory boom in the *Journal of Modern History*, putting its future in historical perspective.[5]

---

[4] See the collaborative study *Memory Unbound*, ed. by Lucy Bond, Stef Craps, and Pieter Vermeulen (New York: Berghahn Books, 2016), which emphasizes the transdisciplinary nature of memory studies today and its diversification into such tributaries as "prosthetic memory," "algorithmic memory, "palempsestic memory," "digital memory ecology, and "petromemory."

[5] Gavriel Rosenfeld, "A Looming Crash or a Soft Landing? Forecasting the Future of the Memory Industry." *Journal of Modern History* 81 (2009): 122–158.

He couches his argument in terms of the rise and fall of its topical interest. He poses the question: will the boom crash as precipitously as it had come into being? Or will interest fade away gradually, marginalized or integrated into other approaches to cultural history? He chooses neither one nor the other scenario, but rather uses the speculative question as a basis for forecasting what the future may hold in light of all that has been learned over the course of three decades of scholarship. He makes an initial distinction between what he characterizes as the "memory boom" and the "memory industry." The former refers to the emergence by the 1980s of historiographical controversies over how the past should be remembered. The latter concerns all the scholarship that these controversies have engendered and that place them in a larger social and cultural context. Rosenfeld relates these historiographical trends closely to leading events in the history of the late twentieth century.

The boom, Rosenfeld argues, resulted from nagging reminders of "unmastered pasts," stemming from World War II. After the war, unresolved issues about its legacies were set aside in the interest of moving on to the tasks of reconstruction of both the nation-state and the worldwide community of nations. Planning for the future loomed large, and meditation on the meaning of the death and destruction wrought by war was held in abeyance. The genocide of European Jews was the most salient among these unresolved historical issues, but there were others; the Japanese atrocities of occupation in China, the American use of atomic weapons, and the violence attending the reluctance of European nations to the demands for independence by their African and Asian colonies. The willingness of the American baby boomer generation of the 1960s to confront issues of racial discrimination, women's rights, and the inequitable distribution of wealth among nations prepared the way for revisiting contested issues of World War II not yet addressed. The preconditions for that discussion were set by the failure of 1960s reformers to advance their left-leaning agenda. The transition marked a reorientation from issues of ideology to those of rhetoric—the power of discourse to direct attention to issues dear to particular groups. As Rosenfeld puts it: the "politics of redistribution" gave way to "the politics of recognition." So identity politics came into being during the 1980s, setting the stage for confrontations among historians over how the past ought to be remembered. It was in this context that the Historians' Dispute in West Germany over the meaning of the Holocaust in German history dramatized this unresolved issue. Postmodern theory lent moment to their debate, for it provided a rationale for contesting how the story should be told. The long-standing notion of history as a unified narrative gave way to one that acknowledged a pluralism of approaches, for each group remember its experience in a different way. By the 1990s, a "memory industry" in historical scholarship was well under way.

By that date, Rosenfeld explains, many of the issues of contestation that had brought the topic of memory to the fore had been addressed. Poignantly, some governments, notably the Federal Republic of Germany, acknowledged guilt and responsibility for the crimes of its predecessors and instituted a politics of atonement. Ironically, Rosenfeld points out, the resolution of issues that had made memory such a topic of contestation now contributed to its marginalization. By the turn of the twenty-first century, the memory boom was cresting. Its descent may have been hastened by the 9/11/2001 terrorist attack on symbols of American government and culture. International terrorism jolted historians out of the luxury of their theoretical preoccupations. The hard realities of the new century incited a desire once more to understand the past as objectively as possible, and with it a search for more unity of interpretation in assessing the larger historical meaning of the present age. History was reasserting its claims upon memory.

Rosenfeld also directs attention to the cycle of academic historiography. Even the most profound historiographical movements have a lifespan. The Annales, Frankfurt School Marxism, psychohistory, he points out, all flourished during the twentieth century but eventually receded from the spotlight of scholarly attention. In the face of the deflation of the memory boom, he considers its effects on the memory industry (i.e., scholarship and teaching dedicated to it) within the academic community. Was there any longer a market for the expanding volume of scholarship it was generating? Would the industry have sufficient staying power for graduate students to justify an investment of their time and money in its research?

Memory studies among historians, Rosenfeld concludes, would seem to be drifting toward the margins of scholarship. But their influence is hardly gone. The memory industry, he argues, is too deeply institutionalized to suddenly disappear. It has, moreover, served a purpose that has outlived the boom, for it has revealed that history and memory as strategies for approaching the past are not as opposed as once believed. At times of crisis and controversy, as exemplified in late twentieth-century historiography, their interaction is more visibly revealed. Henceforth, the study of memory has a recognized place in historical scholarship.

Arlette Farge: A Historian's Sentimental Journey
into the Material Archive

Historian Arlette Farge (b 1941) addresses the memory/history issue from another tack. Aware of the historical transformation of the archive that was under way—from a material place of memory into a digital repository in which the nature of the archive was being reconceived—she offers her thoughts on what might be lost in the process: the research aspect of the historian's way of life. She wrote a memoir about her scholarly work, with a certain nostalgia for the material archives in which historians of her generation had conducted their research. She reconstructs with evocative charm the archival milieu in France in which so many historians of her generation were trained in the mid-twentieth century.[6] She conveys her personal impressions as a researcher in French public archives. She makes her case for the task of gathering, collating, and assessing the evidence the archives contain, not merely as groundwork for writing a compelling interpretative narrative, but also for ascertaining with as much certainty as possible the reality of what happened in times past. If history is an art of rhetoric as postmodern theorists maintain, she argues, it remains a science of detective work in establishing the validity of the evidence on which it is based.

Farge published her memoir, *The Allure of the Archives*, in 1989.[7] The appeal of her little book lies in her capacity to convey what everyday life was like for the researcher in the French archive, complete with the emotions they aroused within her. Farge loved the archive for its quaint eccentricities. For those who have never been *habitués* of the French archives, she suggests, the enchantment of these settings might be difficult to imagine. What might be the allure of drafty corridors, impatient clerks, competition for the best seats in what were less than hospital surroundings,

---

[6] Like Farge, I came to love research in the French public archives and libraries for their idiosyncratic ways, ascetic environment, old-world routines, and endless possibilities for original research. In my day, the archive was the historians' home. It was not just the place where documents were housed. It was one where hidden worlds of the past awaited rediscovery. One learned to master its methods and to fathom its resources, if one was to be taken seriously as a scholar. More than once, I have traveled 3000 miles to see rare texts and documents, many of which today I can access instantaneously on my computer. Looking back, it was in the difficulties of research in those days that contributed to the mystique this place of memory held for me.

[7] Arlette Farge, *The Allure of the Archives* (1989; New Haven, CT: Yale University Press, 2013).

notably in comparison with sitting in one's own comfortable office and retrieving documents online? Farge's point is that for historians in the now vanishing modern age of scholarship the archive served as the tangible place for a way of life valued above all else as the premier repository of the documentary evidence indispensable for serious scholarly research. The material archive, with its vast holdings of manuscripts, newspapers, and rare books, held the resources that made possible the fathoming of an otherwise inaccessible past. For all of its inconveniences, scholarly work in the archives provided priceless opportunity for research and writing about a past whose realities had over time been worn into generalized remembrances that were at best half-truths. Insofar as possible, the historians' high calling was to reenter that past to investigate its complexities on the basis of tangible evidence and so to write interpretive studies that came closer to the truth about what those times had been.

The measure of success in accomplishing this task well, Farge argues, was the historian's willingness to pursue archival evidence with patient persistence. It was a precondition of the depth of the historian's interpretative grasp of the evidence surveyed. But there was always more to do, for the material archive is a resource of cultural memory that can never be completely mastered. In this way, Farge would tug us back from the temptation of relativism implicit in the rhetorical turn of postmodern scholarship. For her, the allure of the archives lies less in the writing, more in the search for evidence, which in its own way could be just as elusive. Data does not speak for itself, Farge explains. Their meaning must be drawn forth through an enlightened engagement by historians with this material resource. The tangible reality of the material archive in the age of scholarship now passing was the guarantor that such a task could be conducted with integrity in the pursuit of widening our knowledge of the past.

Farge's interest was urban life in eighteenth-century France, particularly as lived on the margins; her primary research resource was the archives of the Prefecture of Police of Paris. Exotic characters out of the past populated the police reports that she was reading. The present-day researchers who populated the seats at the archives where she worked were exotic too. They shared a common task, as an imagined community. Farge's scholarly work was in collective mentalities, much in vogue during the 1980s. More specifically it concerned everyday life in the eighteenth-century French city, with particular attention to delinquency. The police archives, especially those of Paris, were ideal for her research. There had been a tendency in the study of mentalities to look for common denominators in the speech and manners of a given population—the ways in which ordinary people shared homogeneous attitudes and values. There was a common sense of social norms, Farge points out. But its nature was discerned at the margins where norms were violated and the boundaries of social conventions were transgressed. In the police archives, the study of mentalities became an exciting field of study for the tensions, confrontations, protest, and defiance one found there. The "mentality" of those times was drawn forth out of this heterogeneous mix, and the historian's skill in doing so was all important. One took no statement at face value. One circled the milieu and surveyed the petty crimes and misdemeanors of everyday life in the city from varied perspectives. The way toward understanding what that society was like was to be found in the interplay between the individual's quest for self-expression and the social conventions that bound the community together. The police were arbiters caught in between. Hence the value of the reports they wrote then for historians now, who collate, arrange, and interpret their meaning from a critical distance and with a different purpose in mind.

Farge explains that the police reports of these judicial archives opened windows upon the everyday life of ordinary people in the city. They highlighted personal conflicts, adversarial relationships, and venal acts, all charged with high emotion. In the police tribunals, delinquents sparred with their interrogators. The police archives were a remarkable source of evidence, for they provided testimony not only of legal rules but also of social norms in the eighteenth century. But the meaning of these documents was not transparent. The historian's task was to make sense of them.[8] The reports had to be interrogated for the hidden meanings that might be culled from them. The interpretive possibilities were endless. But do not let expectations run free of realistic assessments, she cautions. Documents may be tangible evidence of what transpired and what was said. But the closer one looked at them, the more certainties about their meaning vanished.

The year 1989, the date of publication of Farge's memoir, is significant as a symbolic marker in the long-range transition from print to digital communication. While it was already apparent that the digital

[8] For further reflection on this topic, see also the interesting study by Carolyn Steedman, *Dust; The Archive and Cultural History* (New Brunswick, NJ: Rutgers University Press, 2002), which explores the relationship between evidence and representation in historical writing.

revolution in communication was gathering speed, its transforming effects upon research and scholarship had not yet come under her
purview, or at least were not yet her concern. Her memoir might be
regarded as a valedictory salute to historiography as practiced in an age
in which scholarship was so closely identified with the protocols of print
culture. The historian worked with unedited documents, mostly manuscripts. The archives' appeal as repository of reliable evidence lay in the
material tangibility of the documents themselves. The historian's task
was to make sense of them. Here Farge displays her methodological
sophistication. Documents, she explains, do not reveal their meaning
transparently. The closer one looks at a document, the more its meaning
is unsettled. Documents must be interrogated for the knowledge they
may disclose. Archival documents are of all sorts. Even when classified in
preliminary ways, their form and content can be haphazard. Yet the truth
about the past resides therein and the meanings the historian may draw
from them are inexhaustible. The possibilities for interpretation are endless and even the most assiduous researcher cannot hope but to survey a
modest portion of its holdings.

Though steeped in the old ways of evidence gathering, Farge is sensitive to the new methods of language analysis developed in the postmodern rhetorical turn. An event, she explains, is defined out of the interplay
between words and deeds. She cites Foucault on his contribution to our
understanding of discursive practices, and she notes the value of "thick
description" analysis devised by the anthropologists of her generation.
By studying these documents in their ensemble, the researcher learns to
discern the gist of social norms and the thresholds of their transgression.
The task is daunting because the documentary record is at once vast
yet fragmentary. Individual reports fell into the dossiers of police fil-
ings haphazardly. They were terse, composed of clipped narratives, often
incomplete. In and of themselves they tell no complete story. One has
to understand the milieu in which they were offered, notably the bias of
the police officers, and insofar as possible, the mindset of the delinquents
about whom they reported. The researcher has to interpolate the written
words for the hidden attitudes and tacit understandings they presuppose.
Like the design of a broken mosaic, one can decode attitudes prevalent in
this milieu from the parts given in the reports and so complete the pattern. The meaning of this cultural world is to be grasped in the plausible
or the probable as often as in the certain. But in studying the circulation

of ideas in these encounters between police and populace, one can come to understand the nature of social interactions in a cultural milieu so different from our own. Most important in this research, she argues, was discerning the way public dialogue straddled the boundary between orality and literacy. In eighteenth-century Paris, Farge allows, most delinquents, indeed most ordinary people, had limited education, and expressed themselves in an array of verbal skills that owed more to orality than to literacy. Most of the people involved in an incident were not writers, some of them not even readers. The police reports were of value for having "captured the spoken word" in the testimonies they recorded. Among other hints, one looked for the rhythmic inflections of the spoken word. Through their written record, they opened access to the common world of everyday orality in the marketplace and other public venues. In the terminology of Jan Assmann, the police archives were repositories of cultural memory that permitted entry into communicative memory. In the "frozen speech" of cultural memory, communicative memory is preserved. Captured speech was suspended in time, a moment preserved for the historian's attention. The archives held countless moments of that sort. The document may be tangible and comparatively stable. But the world it reveals was dynamic, fraught with tension and ambiguity.

Farge's memoir, therefore, provides insight into what is lost as the archive itself is lifted out of material culture into the cyberspace of digital communication. The original document, however fragmentary and seemingly innocuous, possesses an aura that is lost when it is transformed into a digital simulacrum.

Robert Darnton: Saving the Book in the Age of the Internet

Today we wonder whether print culture is dwindling into insignificance, given the overwhelming influence of media culture. The invention of the World Wide Web has vastly accelerated the long-range trend toward the externalization of memory's resources for preservation and retrieval of information, further obviating the need for a well-ordered mind for data recall. The display of powers of rote memory through the memorization of poetry and apt quotations has long since lost the esteemed place it once held in pedagogy, and has been relegated to televised game shows as an impressive if inessential talent in a culture in which so much information is immediately available online.[9] But digital technology has also come to provide stimulating forums for unleashing memory's imaginative powers to fashion imagery in a newly created cyberspace in which the boundary between memory and fantasy is easily traversed.[10] Historian Robert Darnton (b 1939) also addresses the implications of digital archiving for historical research. He offers a savvy interpretation of the heralded values of the digital archive in light of what might be lost in discarding the tangible record of the past. At the same time, he is actively engaged in saving the important legacies of print culture so that they may live on in the digital age.

In 2007, in the face of the onslaught of the digital age, Darnton decamped from Princeton to take a post at Harvard as head librarian. A student of media in the day of the democratization of print culture, he was concerned for its fate in its waning days. He was well-known among scholars for his history of the production of the *Encyclopédie* as the masterwork for the organization of knowledge during the Enlightenment, and among the public at large for his best-selling *The Great Cat Massacre*, vignettes of the experience of participants at various stages in the writing, production, and dissemination of print culture.[11] Those studies concerned the invention of that culture; now he worked to further the possibilities of its preservation. Historian in a world of librarians, he took up the practical task of figuring out how the integrity of the archives of manuscript and print culture might be preserved in the fast-moving field of the digital remediation of the literary heritage of Western culture. His interest was in the best means of preserving the material sources of research, as well as their transfer into digital archives. Books, he contends, should remain as islands of our print heritage within the sea of new media.

Darnton deeply appreciated the openness of the eighteenth-century republic of letters. He wanted the digital archive to remain the same. He championed the need for open access to books in their digital format, just as the free public library had been the pride of the democratization of

[9] On present-day efforts to redeploy the practice of the art of memory as a mind-game exercise, see the memoir by Joshua Foer, *Moonwalking with Einstein; The Art and Science of Remembering Everything* (New York: Penguin, 2011).

[10] Jay David Bolter and Richard Grusin, *Remediation; Understanding New Media* (Cambridge: MIT Press, 2000), esp. 53–84.

[11] Robert Darnton, *The Great Cat Massacre and Other Episodes in French History* (New York: Basic, 1984); idem, *The Business of the Enlightenment: A Publishing History of the Encyclopédie, 1775–1800* (Cambridge: Harvard University Press, 1979).

print literacy. Ideas should circulate freely and the archive should be open to everyone. Darnton is insightful on the implications of the transition to digital archiving and online publication. But he holds fast to a bias for the ongoing usefulness of printed books. Books possess properties that cannot be replicated. They are tactile, permit flexible consultation, and easy marginal notation. He worries about the danger of knowledge lost, were old books to be discarded as irrelevant. It is in their pages that historians sometimes find their greatest treasures for explaining the nature and meaning of the past.

Over the course of the past decade, Darnton has written a series of articles concerning the role of the book within the digital age.[12] He remains a passionate advocate of the enduring importance of the book in print form. At the same time, he is keenly attuned to the revolution in electronic communication in our times and embraces its possibilities for enriching contemporary culture. No one, he argues, would dispute its transformative power in reshaping cultural memory. But he defends the ongoing value of the book in the midst of this transition. Harvard was then in the midst of negotiation with Google, the ambitious search engine entrepreneur, for digitizing its vast material holdings. Darnton saw his role as a well-timed opportunity to play a role in guiding the move of book culture into the digital age, one analogous to that of scholars who ushered in the republic of letters two centuries before. As a student of the history of the book, he believed deeply in the cause of the philosophers of the Enlightenment for the free and open dissemination of knowledge. He wanted to insure that the digital revolution would proceed in accordance with that high ideal. Google may have been a commercial enterprise. But its directors professed to favor open access to the information their search engines would permit. In his enthusiasm for this cooperative venture between private and public enterprise, he gave Google the benefit of the doubt about what that alliance might be. A few years into this experiment, however, he conceded that he had been overly optimistic in his expectations of harmony in this venture. Copyright law was an impediment. There was also the danger that Google Search would become a monopoly. There were issues concerning the enormity of the task. Even in transcribing ten million books, Google would have copied only some 40% of all those in existence. Digital

---

[12] In the *New York Review of Books*, and collected here as *The Case for Books; Past, Present, and Future* (New York: Public Affairs, 2009).

technology, moreover, would reconfigure the formatting of books, with far-reaching implications for the way they are preserved and read.[13]

At the same time, Darnton wanted to put claims for digital communication as a transformative medium in perspective. As historian he took the long view, for he saw the coming of digital communication as but one more step in the long history of inventions in the technologies of communication. He therefore interpreted this transition, as significant as it might be, from the perspective of continuity with like transitions dating from antiquity. Some of these had remarkable staying power. He noted the invention of the codex to replace the scroll in the third century CE as a boon to the dissemination of Christianity. So facile was the codex as an instrument for reading that it has survived into our own times. He noted as well that the technology of moveable type, invented by Johannes Gutenberg in the mid fifteenth century, had changed little over the following four centuries. What changed during that period was not methods of production of print matter but rather its ever expanding influence, thanks to the slow and steady democratization of literacy from the eighteenth century. Thenceforth, the reading public grew by leaps and bounds. In other words, it took time for the full implications of the coming of print culture to be realized. While today we may wonder at the accelerating pace of change wrought by digital communication, it will take time to understand many of its cultural consequences.[14]

Given his commitment as historian to the integrity of archival research, Darnton has been particularly concerned with the preservation of books and newspapers as material records as the cultural memory they preserve pass into the medium of cyberspace. In digital formatting, much can be excised; even more rearranged. Data threatens to be lost, obscured, or discarded in the process. But he points out that the coming of digital technology, with its dangers for destabilizing the archive of knowledge, has made us aware that the material text of the age of print culture was never as stable as we imagine it to have been. The book as material artifact was full of inconsistencies in the manner of its production. Recent research on the publishing industry in the eighteenth century, he notes, revealed considerable variation in the production of books in that era. There was a sequential order in the production of the book—from compositor to printer to bookseller to reader. But the process was often altered along the

---

[13] Darnton, "Google and the Future of Books," ibid., 1–20.
[14] Darnton, "The Information Landscape," ibid., 21–41.

way, rendering problematic the notion of a standard edition, particularly in the early years of print culture. Interventions were possible at each stage of the process. For the scholar, ascertaining the nature of the original edition was a painstaking task, whose limits are exemplified in efforts to reconstruct the transcription of Shakespeare's dramas. Such scholarship is recondite, yet essential in the integrity of the pursuit of the truth about the past.[15]

Darnton also addressed the closely allied issue of changing reading habits. People in the seventeenth and eighteenth century read differently from the way we do today, he explains. They did not peruse the page in a linear fashion from beginning to end but rather looked for signifi-cant passages here and there to which they responded personally. They copied these into what were known as commonplace books, which as a literary genre might be said to lie somewhere between a diary and a set of reading notes. The readers' intent was to personalize their reading—to extract passages from the text that responded to or reinforced the way they understood the world. They would then integrate these passages into their own conceptual framework. Authors and their ideas were considered instrumental rather than original, bearers of cultural memory that readers would integrate into their personal memories. These commonplace books, therefore, have proven valuable for insight into the mentality of readers in that era. Reading arrived at a middle ground between rote copying and critical interpretation, albeit from a highly personalized point of view. In the eighteenth and nineteenth century, Darnton explains, the invention of the novel took the next step in tipping the balance once more toward the authors' creative role in envisioning a world apart in which readers can immerse themselves. In this transition from the "letters" of the seventeenth century (highly personalized) into the "literature" of the nineteenth (a canon of knowledge), the cultural memory conveyed through reading matter came to be appreciated on its own terms. Darnton's point is that reading habits are changing again in our digital age as we move from the canon to the internet as our mnemonic frame of reference.[16]

Darnton himself experimented with authoring an e-book for the digital age, which he envisioned would be read in a way that suited the technolo-gies of our times. He persuaded a major foundation to fund his experi-ment. His argument was this: In the modern age of print culture, one read

[15] Darnton, "The Importance of Being Bibliographical," ibid., 131–48.
[16] Darnton, "The Mysteries of Reading," ibid., 149–73.

the book in a linear fashion as a single narrative. Authors composed their texts to be read horizontally. In the digital age, by contrast, the e-book might be composed to be read vertically as an echelon of parallel narratives about the same material, but descending from summary argument into the details of the basic research. The e-book so conceived permits several points of entry into its subject matter, depending upon the depth with which the reader wants to explore it. It includes a basic narrative that everyone will read, but also underlying narratives for those who wish to investigate more complex aspects of the topic. Deeper still are primary references for those interested in the sources. These are the equivalent of footnotes in the book of print culture, though fuller and unconstrained by the limits of space.[17]

Darnton's e-book as pyramid, with in its hierarchy of resources of knowledge, is not unlike Aleida Assmann's pyramid of cultural memory. One proceeds from the apex toward the base of the pyramid, if one seeks deeper knowledge or bibliographical orientation for further research. He thought of his model as a means of encouraging the generation of scholars "born digital" and coming of age to write and publish in this format. The project was slow in getting off the ground. There were fewer applications for foundational support than anticipated. Though these increased, the project ran out of money. Nonetheless, he argues, the venture served as a precedent in this early stage of the book reconceived for the digital age. Like experiments in the Gutenberg galaxy, he believes that the model will eventually come into its own.[18]

Darnton's commitment to open access to information in the digital age—not only for scholars but for readers generally—is a noble ideal that signifies his sense of connection between the democratizing aspirations of the Enlightenment and freedom of communication in our own times. But he is quick to point out that the gap between theory and practice can never be completely bridged. If one worries about the monopolizing power of Google, he points out, commercial incentives have never been absent from the book trade. In the past, it may have been worse. As his review of the bookselling business in the eighteenth century reveals, there was much venality, greed, and crass commercialism that compromised the open access ideal. Finding a viable balance between the needs of commerce and those of the pursuit of knowledge will take time in the present age,

---

[17] Darnton, "E-Books and Old Books," ibid., 76–77
[18] Darnton, "Gutenberg-e," ibid., 79–106.

just as it did back then. One might argue that he also seeks to reinforce the loosening hold that bind print and digital technologies in an age in which memory and history might appear to be going their separate ways.

Josef Yerushalmi on History vis-à-vis Historical Remembrance

Memory studies have taken off as an interdisciplinary field with claims to autonomy of its own. But what of the fortunes of history in light of all that was learned from its encounter with memory? Here we return to the place where we began—the rising awareness of the problematic relationship between memory and history during the crucial decade of the 1980s. Two scholars were prominent in those early reflections: the French historian Pierre Nora, to whom we have devoted a chapter, and the American historian Josef Yerushalmi (1932–2009), who addressed the same issues at the same time, but in the broader perspective of an ancient religious tradition. Nora's large-scale project, *Les Lieux de mémoire*, appeared between 1984 and 1992. Yerushalmi's *Zakhor* (in Hebrew "Remember") was published in 1988.[19] Slim by comparison with Nora's magnum opus, Yerushalmi developed his views as a series of lectures. Both were projects of deep learned reflection and in many ways they addressed the same problem: the waning influence of a tradition of historical remembrance. Like Nora, Yerushalmi's interest in memory emerged as a response to a perceived crisis of historiography. They had a chance to interact personally in France in 1984, when Yerushalmi was invited to lecture at the Ecole des Hautes Etudes en Sciences Sociales.[20]

Yerushalmi was a rabbi and a historian.[21] His scholarship was primarily devoted to the history of the Jews in the Iberian Peninsula during the sixteenth century, a time of persecution yet one that contributed to an awakening of modern Jewish historical consciousness. As a research scholar of the history of the Jews, he found himself caught between his role as historian and his cultural identity as a Jew. In a way, he writes an *égo-histoire*, for he saw himself personally involved in the dilemma he

---

[19] Yosef Yerushalmi, *Zakhor; Jewish History and Jewish Memory* (Seattle: University of Washington Press, 1996).

[20] Ibid., xxix.

[21] See the collection of his essays in *The Faith of Fallen Jews*, ed. David Myers and Alexander Kaye (Waltham MA: Brandeis University Press, 2014).

would interpret. Guiding his interest in the memory/history puzzle was the tension between religious remembrance envisioned as prophecy vis-à-vis humanist history as practiced by professional historians. For Yerushalmi the dilemma was personal. His reflections are imbued with nostalgia over the fading of what was most profound in the tradition out of which he came—as he put it—the "universal and ever growing dichotomy" between Judaism as history and as faith.[22]

Both Nora and Yerushalmi studied the way nations naively sustain traditions of historical remembrance: for Nora republican France's pride and patriotism that bore many of the tenets of civic religion; for Yerushalmi the religious foundations of Jewish identity as an imagined community. For Nora, that tradition was comparatively young, a historiographical tradition born in the incipient Third French Republic in the late nineteenth century. That tradition, Nora explained, was intertwined with the dedication to the civic purposes of the nation-state, prominent enough in its demands for allegiance to be considered a cult of collective identity. The Republic proclaimed the virtues of civic duty. It institutionalized its tenets in public education, commemorative ceremonies, and most profoundly in the writing of a patriotic national history. Nora's project was predicated on the awareness that the cult of nationalism no longer spoke to the realities of late twentieth-century French society. It set him on his journey to probe the deep sources of the French national memory, which he traced as far back as the myriad cultural leavings of the Middle Ages.

For Yerushalmi, the past that he would probe was much deeper, harking back to the coalescence of Jewish identity some 3000 years before. Nora had proceeded genealogically in his search for the sources of French identity. Yerushalmi, by contrast, returned to the origins of Jewish awareness of their identity and reconstructed their understanding of its formation moving forward in time. His crucial point was this: In the beginning, Jewish understanding of memory and history were as one. Jewish identity was based on a religious tradition of historical awareness.[23]

As a people, Yerushalmi explains, the ancient Hebrews forged a cultural identity unique in the ancient world for the meaning they found in history. By this he means not any history, but history as divine revelation of their origins as a people, their exceptional relationship with God, their suffering in their nomadic journey, and their promised destiny of redemption.

[22] Yerushalmi, *Zakhor*, 93.
[23] Ibid., 5–26.

That history was written not by historians but by prophets in what was to become the Pentateuch, the core of the Hebrew Bible. This narrative centered on their exodus from Egypt, wanderings in the Sinai, and resettlement in Canaan, idealized as the promised land. In these primordial times, the Hebrews formed a self-conscious cultural identity. Though they valued history, their understanding of it was theist. They believed that God set the course of this history and periodically intervened in their affairs with signs of his favor or disfavor about their way of life. With their arrival in Canaan, their historical experience was sealed as a basis for religious reflection. The Hebrew Bible became their canon. It set forth the moral laws and ritual obligations by which God expected them to live. Most imposing of all was his overarching command to remember this history and to remain faithful to its teachings. This history of origins, deep intertwined with memory, was the only history that mattered.[24]

Through all their trials over following centuries, Yerushalmi explains, Jews maintained their collective identity by their dedication to a religious conception of themselves as an imagined community tightly bound by practices of historical remembrance reiterating the revelation of God's plan in the time and place of their beginnings. Yerushalmi underscores how striking is the contrast between how much meaning they drew from this brief time in their history and how little from all the time that followed. For nearly 2000 years, Yerushalmi explains, the Jewish understanding of historical time was immobilized, its gaze concentrated on the formative period in which their collective identity coalesced. What was added over time was ongoing interpretation of its meaning as a frame of reference with which to explain present realities. The post-Exodus era witnessed not the writing of more prophetic history, but rather burgeoning commentary on a view of history given canonical status in sacred Scripture. Meaning to be found in the realities of a following profane history was barely acknowledged through the Middle Ages and even into modern times.[25]

As long as post-Exodus history could be bracketed as time without meaning, Yerushalmi continues, the sacred time of their origins sufficed to sustain Jews in their efforts to preserve a sense of collective identity. This is all the more remarkable in that shortly after the canonization of the Hebrew Bible by the first century BCE, Jews in the face of Roman persecution scattered throughout the Roman world. No other ancient

---

[24] Ibid., 21, 44–45.
[25] Ibid., 39–40.

people withstood assimilation to a dominant culture with such steadfast resolve. Still, profane history could not be held at bay forever. Gradually new historical experience in a larger world intruded and it became more difficult for Jews to deny cultural influences from which they had shielded themselves for so long. Amidst rising historical awareness in the modern world, Jews were exposed to secular influences, mixed more freely with people of other cultural backgrounds, and came themselves to prize the principle of religious toleration. In the face of Jewish persecution in Spain in the sixteenth century, some Jewish scholars turned to humanist historical interpretation to explain their predicament. For European Jews, the Enlightenment of the eighteenth century permitted the emergence of a new spirit of mutual understanding between Jews and Gentiles. But it was not until the nineteenth century that Jewish scholars engaged in humanist historical writing in earnest.[26]

Yerushalmi views the nineteenth century as a golden age of European historiography, inspired by and closely bound to the new civic religion of nationalism. Jews had always thought of themselves as a nation in their religious exceptionalism. Now, however, they were called upon to integrate into the secular cultures sponsored by the nation-states of Europe, and were asked to think of their Jewish identity in narrowly defined spiritual terms. On the margins of Jewish cultural life, some scholars addressed their circumstances in the modern world. This brought into being an encounter of two kinds of history, one ancient and theist and the other modern and humanist. Since then, Yerushalmi contends, both have come to exist in uneasy alliance, for the growth of humanist Jewish history paralleled the gradual erosion of religious faith founded on tradition-bound historical remembrance.[27]

Modern historiography, Yerushalmi concludes, became a subversive force, offering itself as a faith for non-observant Jews. As humanist interpretation, it aspired to provide consolation for understanding their time and place in modern history. But this was an unsatisfactory substitute for what Jewish historical remembrance had been, and fell short in any effort to rethink Jewish identity. The cultural force of Judaism lay in the power of historical remembrance to provide coherence and commitment to a way of life through practices that confirmed and reinforced collective identity in ways no secular historiography has and probably cannot. One critic

---

[26] Ibid., 57–74, 84.
[27] Ibid., 86, 95.

claimed that history could repair the broken myth of an ancient way of religious understanding. But Yerushalmi concludes that the rise of the humanist historian within Jewish culture in our times has been "at best pathologist, hardly a physician."[28] History cannot provide what historical memory once nurtured and sustained—a collective identity that had withstood all the countervailing forces that had assailed Jews through the vicissitudes of their fortunes across two millenia. Yerushalmi finds unconvincing the positivist notion of devoting oneself to the study of history "for its own sake," for historical interpretation in its humanist as in its theist guise aspires to transcend mere information in its quest for meaning.[29] There is great modern literature in which Jewish identity finds fresh interpretation for our times, he acknowledged. But such insights are piecemeal and limited in their capacity to convey a sense of transcendence. The prospect of writing a new metanarrative of Jewish history for our times, he concludes, seems a remote and tenuous expectation.[30] Yerushalmi finds consolation in the integrity of his more modest endeavor as historian considered as a way of truth for the humanist knowledge it can provide. Religious tradition continues to play a powerful social role, even though its theology of history can no longer be accepted in a spirit of uncritical faith. Yet there is a creative interplay between the two, for memory and history have different vocations—the memorialist's duty to remember alongside the integrity of the historian's quest to seek the truth about the past.[31]

Paul Ricoeur on History and Memory

Yerushalmi's study invites comparison with the last meditation of the late French philosopher Paul Ricoeur (1913–2005), who configures the relationship between memory and history from the vantage point of our contemporary concerns. His study *L'Histoire, la mémoire, l'oubli* (2000) provides a comprehensive phenomenology of the place of memory in contemporary historiography, and in its way visits the major issues that we have addressed in this essay: orality/literacy, commemoration, trauma.[32]

---

[28] Ibid., 94–98.

[29] Ibid., 100.

[30] Ibid., 99–101.

[31] See the Nora/Yerushalmi comparison by Susan Crane, "Writing the Individual Back into Collective Memory," *American Historical Review* 105 (December 1997): 1380–1383.

[32] Paul Ricoeur, *La Mémoire, l'histoire, l'oubli* (Paris: Editions du Seuil, 2000), 3–4, 167–180, 642–656.

Memory, Ricoeur explains, informs history in so many ways. History appropriates the testimony of memory and then refashions it. It benefits from memory's animating spark. It aspires to draw close to memory's capacity to evoke the imagination and to touch those emotions that make the past "live again." Ricoeur has much sympathy for, or at least understanding of, the historian's wish to make real the impossible dream of "resurrecting" the past. He notes that it is a desire that reaches far beyond Jules Michelet's romantic fantasy, for historians aspire to be more than undertakers of the dead. "Is it not the ambition of every historian," he asks, "to uncover behind the death mask the face of those who formerly existed, who acted and suffered, and who were keeping the promises they left unfulfilled?"[33]

While acknowledging the tight relationship between memory and history, he wished to point out some differences in their resources, and so to reclaim for memory its autonomy. History in its quest for the truth of the past, he explains, may be more encompassing than memory. It possesses a wider array of resources and it "layers time differently." It may elaborate on the meaning of the past in useful conceptual structures that are its own invention. Living memory is fleeting; history more enduring. But memory cannot be reduced to the status of a mere object of the historians' scrutiny. Memory has a different vocation. Whereas the vocation of history is to search for the truth of the past, that of memory is to safeguard the experience it evokes. Memory's greatest asset is its care (*souci*) for the past, even if it can only hold our attention for so long. Memory is distinguished from history in its act of recognition, a category of understanding for which history has no equivalent. For this reason, memory is a "little miracle," for it may evoke the presence of the past suddenly and randomly. Recollection brings with it the energy of inspiration. Memory may be fleeting. Yet it animates the moment, and for this reason may quicken the historian's interest. Ricoeur labels this moment of recognition "happy memory." It reconciles the absent past with the living present. Memory may fade. But with its rediscovery, one may care again. The past lives once more in memory's animation.

To explain what is at stake, Ricoeur journeys back into those primordial depths out of which the distinction between memory and history initially emerged. He locates that parting of the ways at the threshold at which orality yielded a place to literacy. He bases his analysis on the first philo-

[33] Ibid., 649.

sophical effort to address its implications for the uses of memory: Plato's Socratic dialogue *Phaedrus* (circa 370 BCE). It is the most ancient reflection on the puzzle of memory's relationship to history, here couched in a conversation about the effects of writing on the human imagination.[34] Socrates poses the question: is writing a remedy or a poison as an aid to memory? His answer lies not in choosing one or the other, but rather in considering how the invention of a new technology of communication transforms our understanding of memory's resources. Memory evokes the past in its transcending imagination; history circumscribes the past within its interpretative framework. History is thus a remedy for the instability of the evanescent human imagination. But it is a poison insofar as it forestalls other ways in which the past might be envisioned. History, Ricoeur explains, may be counted among the first arts of memory. It holds the past fast, memory in its preservationist mode. History, therefore, may be regarded as a kind of commemoration. It is our most informed route toward a critical understanding of the past, but comes at the price of conferring determining limitations on what and how we shall remember. Memory in its mode as imagination, by contrast, must be treasured as an unbounded power of mind. In its inspiration, it emboldens new beginnings. In acknowledging the paradox of its faithfulness to the remembrance of inhumanities out of the past—forgiving yet never ceasing to care all the while—memory may be reconciled with the bleakest historical realities. Here memory quickens out of history's record, a resource that sustains our hope for the redemption of an errant past.

For Ricoeur, history, for all its power to pursue the truth about the past, possesses no autonomy vis-à-vis memory. It may be regarded as one among the arts of memory. In the act of writing, history represents the past for the present, and so establishes critical distance from it. Its interpretative framework invokes a method for selecting what is worthy of remembrance and what is to be forgotten, and it has staying power. It establishes a framework based upon concrete evidence that stabilizes the recollection of the past.

---

[34] Ricoeur's interpretation is based on his reading of Jacques Derrida's meditation on "Plato's Pharmacy" in his *Dissémination* (Paris: Seuil, 1972), 71–197.

How Interest in Memory has Influenced Our Understanding of History: Some Conclusions

It all began as a crisis in French historiography in the late twentieth century, as historians sensed that the old narratives of national history no longer spoke to contemporary globalizing realities. Modern history had augured the future in light of a conception of progress, and as such a reference for interpreting the past. Following the initiative of French historian Pierre Nora, historians instead took the present age as their frame of reference and sought to deconstruct the sources of modern French history by tracing them genealogically to places of memory that had given identity to the French cultural heritage. Nora's premise had been that memory and history had long been bound in unknowing alliance, and that these places of memory had over time become barren, "like seashells on the seashore from which the tide of living memory has retreated."[35] His approach had been autopsy. But the influence of his project soon escaped his limited objectives to become a fillip for revitalizing history in a myriad of ways.

The historians' study of memory in the late twentieth century revealed its profound ambiguity, for memory has a creative as well as a preservationist side. Old memories can quicken when interrogated in new ways. What began as autopsy ended in inspiration, as scholars not only in history but across the curriculum turned to the study of collective memory as the sinew of cultural heritage. By the 1990s these avenues of scholarship had converged in what I have referred to as the memory phenomenon. It contributed in rich and powerful ways to the scholarly understanding of cultural history, in vogue among historians since the 1960s but now exploding in research that lent new complexity to the field, and to a rethinking of historiography in the process. Here are a few of the ways I have addressed these topics:

The Politics of Commemoration

In the early twentieth century, the historians' interest in collective memory had directed attention to immemorial tradition—the weight of custom, the inertial power of habits of mind, social traditions as they

[35] Pierre Nora, "Entre Mémoire et histoire," in *Les Lieux de mémoire* (Paris: Gallimard, 1984), 1: xxiv.

served as bulwarks against the tide of innovation and rapid change. That discourse about the nature of tradition was challenged during the early 1980s by historians Eric Hobsbawm and Terence Ranger in their argument about the invention of national traditions. It was a first step in the historians' reflections on the dynamics of cultural memory. Collective memory, once judged imponderable, was recast as volatile—present-minded in its objectives, fragile in its susceptibility to political manipulation. Tradition once viewed as the common sense of community was reconsidered for the way in which it was contested over time. The legacy of commemoration lay not in the authority of the past but the politics of the present.

Cultural Communication

The revolution in the technologies of communication in the late twentieth century set historians thinking about like inventions in the past, notably that of manuscript literacy deep in antiquity, and of print literacy in the early modern era. With each revolution, the nature of collective memory was reconceived. Studies of collective memory at these thresholds revealed how different were the mindset on either side of these crossings. This avenue of inquiry provided a plotline for a history of the cultural effects of advances in the technologies of communication from antiquity to the present. Orality/literacy was the first threshold to attract the historians' attention, as they developed techniques for reaching into cultures of primary orality by ferreting out oral residues within manuscript texts. Scholarship on print literacy followed, with particular attention to the spread of print literature and the democratization of reading during the Enlightenment of the eighteenth century. These developments permitted the formulation of more abstract conceptions of collective memory and hence of social identity. The idea of a "republic of letters" among educated people during that era anticipated the idea of "imagined communities" by the end of the nineteenth century. The transition into digital age technologies is the site of research currently under way, one that promises to guide the future of memory studies.

Studies in cultural communication also deepened understanding of the channels through which cultural memory is transmitted. There is a predictable logic to the ways of transition over time: from witness, to hearsay, to narrative clusters, to abstract idealization. But this interpretation of collective memory's cycle is normative, not deterministic. Collective memory is fragile and protean, and the pattern of its transmission is easily disrupted by unanticipated contingencies and in any case distorted by the changing realities to which it adapts. Collective memory moves with the times.

The Disabling Effects of Trauma Upon Historical Understanding

Studies of collective memory also reveal how trauma can repress memory of deeply wounding psychological experience and so retard accurate historical assessment of the realities of the past. This avenue of inquiry into collective memory is of a different order. Here the problem is not idealization of a consciously recognized memory, but redemption of one that remains to some degree hidden in repression. Repressed memory blocks historicization, demanding acknowledgment of suffering and injustice and enjoining atonement before the historian may proceed. In these circumstances, the historian must persuade memory to free up its secrets. As Paul Ricoeur noted, the unrequited memory of trauma defies closure. The historian must first beg pardon for a wound that cannot yet be closed. This became the painful burden of decades of efforts to come to terms with the Holocaust. In the face of the impending passing of the last Holocaust survivors, a younger generation of scholars was called upon to assume the added responsibility of deciding how the living memory of the Holocaust memory would pass into history.

As the calculated genocide of millions of innocent victims, the Holocaust defies even complex interpretation and there are grounds for judging it exceptional. Yet historical remembrance never stands still, and, as Daniel Levy and Natan Sznaider have demonstrated, has since been lifted onto a global plane, an atrocity whose study elicits empathy among people of goodwill everywhere. As a crime against humanity, its memory has been enlisted in the cause of advancing universal human rights. For historians, the Holocaust has come to serve as a model for genocides in other times and places, some much earlier (the Terror during the French Revolution, for example), others extending into our own times (Bosnia during the 1990s; Syria in the 2010s). Such studies are now a mainstay of historical research.

History Unbound

In the breakdown of old historiographical frameworks, it was not only memory that would loosen binding ties. The study of memory revealed the constraints under which historical scholarship had long operated. History had been tightly bound to the protocols of print culture. Historians defined their task as one of producing written texts based on manuscripts housed in public and private archives, and they were often naive about the cultural memories that tacitly motivated their work. These were regarded as stable sources of remembrance about which they could report with authority. The historians' archive expanded considerably during the early twentieth century, as they consulted a wider range of research materials. Once seeing their discipline as one apart from the other human sciences, they learned from the subject matter and the methods of related disciplines, notably anthropology, in the development of a new cultural history by the 1960s.

But the mnemonic turn in historiography liberated history from its almost exclusive identification with print culture, as it became possible to use the resources of electronic media to complement the written narrative. In the scholarship of the late twentieth century, memory and history became traveling companions. In studying memory, historians had over time come to acknowledge affinities between memory and history as perspectives on the past. The memory phenomenon revealed how collective memory had tacitly prompted the framing of modern history, even as it played a subordinate role. The lure of memory had always been to make the past "come alive" again, and that notion has long served as the motto of the amateur historian and the history buff. In the midst of the memory phenomenon, "performative history" gained currency among some professional historians as well. In the intensity of the study of memory in the late twentieth century, therefore, history came to be deeply influenced by the object of its inquiry, as historians came to understand how the study of memory was widening their conception of history.

The Expanding Domain of Historiography

Memory studies have contributed to the expanding domain of historiography within the world of scholarship. I note three phases: the concentration on problems of research during the early twentieth century, largely those of finding and evaluating evidence; the linguistic turn of the

1970s, which addressed styles of representing the past in the rhetoric of composition; the mnemonic turn of the 1980s, with new attention to subjective experience in constant tension with representation. The memory phenomenon was to some degree a response to the excesses for the claims of representation upon historical interpretation. It is true that there can be no experience of the past that is not mediated. On the other hand, memory studies provided an opening toward ways of apprehending subjective experience. If historians cannot resurrect past experience directly, they can move closer to it by stimulating the historical imagination through modes of historical reenactment. Performative history acquired a place in contemporary historiography as a consequence.

The memory phenomenon, too, raised in a new context the old historiographical conundrum of cyclical versus linear conceptions of history. It promoted awareness that it was the pattern of historiographical evolution of a field of scholarly interest, not that of history itself, that moves in cycles. The historians' subject matter was born of events, reported as testimony, discerned in pioneering trends of scholarship, articulated in guiding models, gravitated toward normative narratives, before exhausting its potential for discovery and interpretation in increasingly discrete case studies. In this cyclical pattern of addressing a topic, historiography mimicked the cycles of memory: from testimony to storytelling to idealization before moving into abstractions that lost touch with those concrete memories that had once exercised such powers of persuasion.

Historians affirmed a new respect for historiography in light of the pluralism of approaches that the memory phenomenon had unleashed. From the quest for a unified grand narrative of history, they turned to topical places from which to develop particular narratives localized on a spatial map of places not unlike those of the ancient art of memory. As collective memory gravitated from the old places of national memory to new ones of a globalizing culture, history followed.

Mnemohistory as the De Facto Master Narrative of Memory Studies

Tracing these chains of collective memory was the basis for a method Jan Assmann pioneered as "mnemohistory." This method shifted interest from origins to legacies. Together with Aleida Assmann, he formulated a method for studying the history of what has been judged worthy of remembrance in times past, explaining the way it was preserved as it was modeled and remodeled across the ages. In a way, the construction of this storyline in the elaboration of cultural memory would play a role analogous to metahistorical narratives based on linguistic protocols, of the sort earlier developed by theorist Hayden White. Mnemohistory provided the makings of a metanarrative of a different nature.

Mnemohistory as an alternative to conventional historical narrative might be contrasted this way: master narratives represent the patterns of the past in recounting its events; mnemohistory narrates patterns out of the past as revealed in memorable cultural legacies. It singles out what was valued and what was chosen to be preserved. It lends interpretative structure to broadly conceived traditions of remembrance, based upon singular events and remarkable personalities. It provides a new way of understanding cultural history, based on tracing the influence of cultural memory conceived as heritage. Mnemohistory reconstructs chains of remembrance. As an approach to memorable personalities, therefore, interest shifts from the lives to the afterlives of significant historical actors. Scholars study memorable events as they became metonyms, currency in the politics of memory, as, for example, in the symbolic uses of the memory of the Crusades or of the terrorist attack on the USA on 9/11. Here mnemohistory reasserted history's claim to mastery over memory, but not without introducing an element of subjectivity into historical interpretation. The effect of mnemohistory's interrogation of memory has been to highlight the fragility of memory in its protean dynamism; but it also draws attention to the instability of historical interpretation, for in pursuing this approach historians take their cue from traditions of remembrance. As memory fades, history is revised.

History Among the Arts of Memory

The memory phenomenon also encouraged a newfound openness to alternative approaches to the past. It acknowledged that history is one among the many arts of memory. The search for a middle ground engaging history as traditionally conceived and history based on forms of reenactment inspired the hybrid conception of "historical remembrance," which Jay Winter characterized as a creative space between memory and history. Unlike mnemohistory, "historical remembrance" is more orientation than method. Rather than being descriptive, it is suggestive. Events and personalities serve as launching pads for the imaginative re-creation of the past. Its goal is not factual accuracy but verisimilitude in approaches that extends history's reach into the imagination of what life was like back then. I have noted the varieties of such "performative history": the historical novel, historic preservation; museums; theater; tourism to sites of memory; historical reenactment. All are calculated to rehearse the events of the past in imagined repetition. Each of these genres dedicates itself to a kind of time travel between past and present, superseding the historicist way of sequencing the episodes of history that seek to establish a continuum between past and present viewed from a critical distance.

The Era of Memory in Historical Perspective

As François Hartog has explained, different ages have viewed memory differently in light of changing conceptions of historical time. In our own, memory and history have grown closer, eliciting a newfound sympathy for a presentist perspective on historical time. The term "presentism" had not even entered the historian's lexicon a generation ago.[36] It had always been memory's way of understanding time. Presentism as a mode of historical interpretation for some time evinced pessimism about the near past and diminished expectations of the future, marking the contemporary age with a sense of its indeterminacy. Old political narratives had foundered and for a time cultural factors came to the fore. These included globalizing consumerism, the politics of gender, and the rise of media culture. The Cold War was waning, and came to an end in the revolutions of 1989 in Eastern Europe and the collapse of the Soviet Union two years later. It led political scientist Francis Fukuyama to prophesy the coming of a peaceable kingdom under the benign aegis of the USA. It rejected the notion of history "for its own sake," drawing past and future into interpretations that served present needs directly.

An End to the Memory Phenomenon?

How long the memory phenomenon will be with us is an open question. As we enter more deeply into the twenty-first century, the human prospect is likely to be reconsidered in terms of more intimidating, indeed unprecedented, realities, many of them political in nature. The year 1989, the salient symbol of the closing of the modern era was succeeded

[36] Patrick Hutton, "Presentism," in *The New Dictionary of the History of Ideas*, ed. Maryanne Horowitz (Detroit, MI: Thomson/Gale, 2005), 5: 1896–99.

by 11 September 2001, auguring the coming of a new one. One wonders whether the era 1989–2001 will come to be regarded as a time of transition between the two, signifying a crystallization of newly imposing historical realities, some in the making for a long while but now asserting themselves, among them a new global politics; terrorism and endemic regional warfare; the resurgence of religious fundamentalism with a fanatical edge; the immobilization of reform-minded policy in the face of resilient neo-conservatism; the mass migration of people seeking to escape poverty and violence; the degradation of the environment with the possibility of irreversible climate change.[37]

All of these realities will generate new concepts of memory. That work is already in progress. But it may as well prompt historians to return to their earlier vocation as interpreter of events, in search of certainties rather than verisimilitudes in the knowledge of the daunting problems we face. Whatever the future of memory holds for historical scholarship, our understanding of history will never be the same again, thanks to the mnemonic turn. The emergence of a new master narrative amidst all the deconstruction of rhetorical forms of composition and the mnemonic ways of understanding seems improbable. Yet the work of historians remains as vital as ever in its vocation to balance initiatives to recapture the meaning of the past with that of ascertaining as accurately as possible the reality of what actually transpired back then.

---

[37] As for memory itself, an understanding of its nature may soon be radically reconceived in light of emerging developments in artificial intelligence that communications scientists dealing with issues of memory have hardly begun to address. Ray Kurzweil's prophecy of the coming of the Singularity (the merger of biological and artificial intelligence) in 2045 may not arrive quite so soon. But artificial intelligence will shape ideas about memory in ways as yet impossible to foresee.

BIBLIOGRAPHY

Agulhon, Maurice. 1981. *Marianne into battle; Republican imagery and symbolism in France, 1789–1880*. Trans. J. Lloyd. Cambridge: Cambridge University Press.

Anderson, Benedict. 2006. *Imagined communities; Reflections on the origins and spread of nationalism*. London: Verso. Rev ed.

Anderson, Perry. 1998. *The origins of postmodernity*. London: Verso.

Ankersmit, Frank R. 2001. *Historical representation*. Stanford: Stanford University Press.

Ankersmit, Frank R. 2005. *Sublime historical experience*. Stanford: Stanford University Press.

Ankersmit, Frank R. 2013. Historical experience beyond the linguistic turn. In *The sage handbook of historical theory*, ed. Nancy Partner and Sarah Foot, 424–438. London: Sage.

Ariès, Philippe. 1988. L'Histoire des mentalités. In *La Nouvelle Histoire*, ed. Jacques Le Goff and Jacques Le Goff, 167–190. Paris: Editions Complexe.

Ariès, Philippe. 1986. *Le Temps de l'histoire*. Paris: Seuil.

Assmann, Aleida. 2008. Canon and archive. In *Cultural memory studies; An international and interdisciplinary handbook*, ed. Astrid Erll and Ansgar Nünning, 97–107. Berliin: Walter de Gruyter.

Assmann, Aleida. 2011a. *Cultural memory and western civilization; Functions, media, archives*. Cambridge: Cambridge University Press.

Assmann, Jan. 2011b. *Cultural memory and early civilization; Writing, remembrance, and political imagination*. Cambridge: Cambridge University Press.

Assmann, Jan. 1997. *Moses the Egyptian; The memory of Egypt in western monotheism*. Cambridge: Harvard University Press.

© The Editor(s) (if applicable) and The Author(s) 2016
P.H. Hutton, *The Memory Phenomenon in Contemporary Historical Writing*, DOI 10.1057/978-1-137-49466-5

Baldwin, Peter (ed.). 1990. *Reworking the past: Hitler, the Holocaust and the Historians' Debate.* Boston: Beacon Press.

Barber, Benjamin R. 2001. *McWorld vs Jihad.* New York: Ballantine.

Bartel, Diane. 1996. *Historic preservation: Collective memory and historical identity.* New Brunswick: Rutgers University Press.

Barzun, Jacques, and Henry F. Graff. 2004. *The modern researcher,* 6th ed. Florence: Thomson/Wadsworth.

Bechtel, Roger. 2007. *Past performance; American theatre and the historical imagination.* Lewisberg: Bucknell University Press.

Benjamin, Walter. 2003. *Walter Benjamin: Selected writings.* Ed. Howard Eiland and Michael Jennings. 4 Vols. Cambridge: Harvard University Press.

Bernauer, James, and David Rasmussen (eds.). 1988. *The final Foucault.* Cambridge: MIT Press.

Blight, David W. 2011. *American oracle; The Civil War in the civil rights era.* Cambridge: Harvard University Press.

Blight, David W. 2001. *Race and reunion; The Civil War in American memory.* Cambridge: Harvard University Press.

Bloch, Marc. 1925. Mémoire collective, tradition et coutume; A propos d'un livre recent. *Revue de Synthèse Historique* 40: 73–83.

Bloch, Marc. 1993. *Apologie pour l'histoire ou métier d'historien.* Paris: Armand Colin.

Bodner, John. 1992. *Remaking America; Public memory, commemoration, and patriotism in the twentieth century.* Princeton: Princeton University Press.

Bodner, John. 2010. *The "good war" in American memory.* Baltimore: Johns Hopkins University Press.

Bolter, Jay David, and Richard Grusin. 2000. *Remediation; Understanding new media.* Cambridge, MA: MIT Press.

Bond, Lucy, Stef Craps, and Pieter Vermeulen, (eds.) *Memory unbound; tracing the dynamics of memory studies.* New York: Berghahn Books, (2016).

Boutier, Jean, and Dominique Julia (eds.). 1995. *Passés recomposés.* Paris: Editions Autrement.

Boym, Svetlana. 2002. *The future of nostalgia.* New York: Basic Books.

Boym, Svetlana. 2007. Nostalgia and its discontents. *Hedgehog Review* 9: 7–18.

Braudel, Fernand. 1969. *Ecrits sur l'histoire.* Paris: Flammarion.

Brundage, W. Fitzhugh. 2005. *The southern past: A clash of race and memory.* Cambridge: Harvard University Press.

Carr, Nicholas. 2011. *The shallows; What the internet is doing to our brains.* New York: Norton.

Caruth, Cathy. 1996. *Unclaimed experience; Trauma, narrative, and history.* Baltimore: Johns Hopkins University Press.

Clark, Stuart (ed.). 1999. *The Annales school: Critical assessments.* London: Routledge.

Cohen, Margaret. 1993. *Profane illuminations; Walter Benjamin and the Paris of surrealist revolution.* Berkeley: University of California Press.

Conan, Eric, and Henry Rousso. 1996. *Vichy, un passé qui ne passe pas.* Paris: Gallimard.

Confino, Alon. 1997. Collective memory and cultural history: Problems of method. *American Historical Review* 105: 1386–1403.

Confino, Alon. 2006. *Germany as a culture of remembrance; promises and limits of writing history.* Chapel Hill: University of North Carolina Press.

Confino, Alon. 2008. Memory and the history of mentalities. In *Cultural memory studies; An international and interdisciplinary handbook,* ed. Erll Astrid and Nünning Ansgar, 77–84. Berlin: Walter de Gruyter.

Connerton, Paul. 2009. *How modernity forgets.* Cambridge: Cambridge University Press.

Connerton, Paul. 1980. *The tragedy of enlightenment; An essay on the Frankfurt school.* Cambridge: Cambridge University Press.

Connerton, Paul. 1989. *How societies remember.* Cambridge: Cambridge University Press.

Coser, Lewis A. (ed.). 1992. *Maurice Halbwachs; On collective memory.* Chicago: University of Chicago Press.

Crane, Susan A. 1997. Writing the individual back into collective memory. *American Historical Review* 105: 1372–1385.

Crane, Susan A. (ed.). 2000. *Museums and memory.* Stanford: Stanford University Press.

Cubitt, Geoffrey. 2007. *History and memory.* Manchester: Manchester University Press.

Daileader, Philip, and Philip Whalen (eds.). 2010. *French historians, 1900–2000.* Chichester: Wiley-Blackwell.

Darnton, Robert. 2009. *The case for books past, present, and future.* New York: Public Affairs.

Darnton, Robert. 1984. *The great cat massacre and other episodes in French history.* New York: Basic.

Darnton, Robert. 1982. *The literary underground of the old regime.* Cambridge: Harvard University Press.

Davis, Fred. 1979. *Yearning for yesterday; A sociology of nostalgia.* New York: The Free Press.

Derrida, Jacques. 1972. *La Dissémination.* Paris: Seuil.

Derrida, Jacques. 1994. *Specters of Marx; the state of the debt, the work of mourning, and the new international.* Trans. P. Kamuf. London: Routledge.

Di-Capua, Yoav, (ed.) 2015. Trauma and other historians. *Historical Reflections* 41(3).

Dodman, Thomas. 2011. Un Pays pour la colonie; Mourir de nostalgie en Algérie française, 1830–1880. *Annales HSS* 63(3): 743–784.

Dosse, François. 1987. *L'Histoire en miettes: Des Annales à la "nouvelle histoire".* Paris: La Découverte.

Dosse, François. 2011. *Pierre Nora; Homo historicus.* Paris: Perrin.

Douglas, Allen. 2002. *War, memory, and the politics of humor.* Berkeley: University of California Press.

Eisenstein, Elizabeth L. 1979. *The printing press as an agent of change.* 2 Vols. Cambridge: Cambridge University Press.

Elias, Norbert. 1978. *The civilizing process; The development of manners.* New York: Urizen Books.

Eribon, Didier. 1991. *Michel Foucault.* Trans. B. Wing. Cambridge: Harvard University Press.

Erll, Astrid, and Ann Rigney (eds.). 2012. *Mediation, remediation, and the dynamics of cultural memory.* Berlin: Walter De Gruyter.

Erll, Astrid, and Ansgar Nünning (eds.). 2008. *Cultural memory studies; An international and interdisciplinary handbook.* Berlin: Walter de Gruyter.

Erll, Astrid. 2011. *Memory in culture.* Trans. S.B. Young. London: Palgrave Macmillan.

Farge, Arlette. 2013. *The allure of the archives.* Trans. T. Scott-Railton. New Haven: Yale University Press.

Febvre, Lucien. 1992. *Combats pour l'histoire.* Paris: Armand Colin.

Feindt, Gregor, Félix Krawatzek, Daniela Mehler, Friedemann Pestel, and Rieke Trimçev. 2014. Entangled memory: Toward a third wave in memory studies. *History and Theory* 53: 24–44.

Foer, Joshua. 2011. *Moonwalking with Einstein; the art and science of remembering everything.* New York: Penguin.

Foley, John Miles. 1988. *The theory of oral composition.* Bloomington: Indiana University Press.

Foucault, Michel. 1988. *The care of the self.* Trans. Robert Hurley New York: Random House.

Foucault, Michel. 1994. *Dits et écrits: 1954–1988.* Ed. Daniel Defert and François Ewald. 4 Vols. Paris: Gallimard.

Foucault, Michel. 1977. *Language, counter-memory, practice; Selected essays and interviews.* Trans. D. Bouchard, S. Simon. Ithaca: Cornell University Press.

Foucault, Michel. 1972. *The archaeology of knowledge.* Trans. A.M.S. Smith. New York: Harper & Row.

Friedländer, Saul (ed.). 1992. *Probing the limits of representation; Nazism and the "Final Solution".* Cambridge: Harvard University Press.

Friedländer, Saul. 1993. *Memory, history, and the extermination of the Jews of Europe.* Bloomington: Indiana University Press.

Friedländer, Saul. 2007. *The years of extermination: Nazi Germany and the Jews, 1939–1945.* New York: HarperCollins.

Friedländer, Saul. 1979. *When memory comes.* Trans. H.R. Lane. New York: Farrar, Strauss, Giroux.

Friedländer, Saul. 1980. *History and psychoanalysis; An inquiry into the possibilities and limits of psychohistory.* Trans. S. Suleiman. New York: Holmes & Meier.

Fritzsche, Peter. 2002. How nostalgia narrates modernity. In *The work of memory: New directions in the study of German society and culture,* ed. Alon Confino and Peter Fritzsche. Urbana: University of Illinois Press.

Fritzsche, Peter. 2001. Specters of history: On nostalgia, exile, and modernity. *American Historical Review* 106: 1587–1618.

Fritzsche, Peter. 2004. *Stranded in the present; Modern time and the melancholy of history.* Cambridge: Harvard University Press.

Fukuyama, Francis. 1993. *The end of history and the last man.* New York: Avon Books.

Funk, Robert, and Roy Hoover (eds.). 1993. *The five gospels; What did Jesus really say.* New York: Harper Collins.

Furet, François. 1983. Faut-il célébrer le bicentenaire de la Révolution française. *L'Histoire* 52: 71–77.

Furet, François. 1995. *Le Passé d'une illusion.* Paris: Robert Laffont.

Furet, François. 2014. In *Lies, passions and illusions; the democratic imagination in the twentieth century,* ed. Christophe Prochasson. Chicago: University of Chicago Press.

Furet, François. 1978. *Penser la Révolution française.* Paris: Gallimard.

Fussell, Paul. 1975. *The great war in modern memory.* Oxford: Oxford University Press.

Gadamer, Hans-Georg. 1989. *Truth and method.* Cambridge: MIT Press.

Gay, Peter. 1985. *Freud for historians.* Oxford: Oxford University Press.

Gérard, Alice. 1970. *La Révolution française, mythes et interpretations.* Paris: Flammarion.

Gervereau, Laurent. Pourquoi canoniser Pierre Nora? *Le Monde.fr/idées/articles/2001/11/01.*

Gilbert, Felix, and Stephen R. Graubard (eds.). 1972. *Historical studies today.* New York: Norton.

Gillis, John (ed.). 1994. *Commemorations; The politics of national identity.* Princeton: Princeton University Press.

Glassburg, David. 1990. *American historical pageantry; The uses of tradition in the early twentieth century.* Chapel Hill: University of North Carolina Press.

Goldhagen, Daniel Jonah. 1997. *Hitler's willing executioners; ordinary Germans and the Holocaust.* New York: Random House.

Golson, Richard J. (ed.). 1998. *Fascism's return; Scandal, revision, and ideology since 1980.* Lincoln: University of Nebraska Press.

Gombrich, Ernst Hans. 1986. *Aby Warburg; An intellectual biography.* Chicago: University of Chicago Press.

Goody, Jack. 1987. *The interface between the written and the oral.* Cambridge: Cambridge University Press.

Gross, David. 2000. *Lost time; On remembering and forgetting in late modern culture*. Amherst: University of Massachusetts Press.

Gross, David. 1992. *The past in ruins*. Amherst: University of Massachusetts Press.

Leroi-Gourhan, André. 1965. *Le Geste et la parole: la mémoire et les rythmes*. 2 vols. Paris: Albin Michel.

Hadden, R. Lee. 1996. *Reliving the civil war; A Reenactor's handbook*. Mechanicsburg: Stackpole Books.

Halbwachs, Maurice. 1971. *La Topographie légendaire des évangiles en Terre Sainte*. Paris: Press Universitaires de France.

Halbwachs, Maurice. 1975. *Les Cadres sociaux de la mémoire*. New York: Arno Press.

Hamer, Mary. 1993. *Signs of Cleopatra; History, politics, representation*. London: Routledge.

Harlin, David. 1989. Intellectual history and the return of literature. *American Historical Review* 94: 581–609.

Hartog, François. 1996. *Mémoire d'Ulysse; Récits sur la frontier en Grèce ancienne*. Paris: Gallimard.

Hartog, François. 2015. *Regimes of historicity; Presentism and experiences of time*. New York: Columbia University Press.

Hartog, François. 1995. Temps et histoire; 'Comment écrire l'histoire de France?'. *Annales HSS* 50(6): 1219–1236.

Havelock, Eric. 1963. *Preface to Plato*. Cambridge: Harvard University Press.

Heckel, Waldemar, and Lawrence A. Tritle (eds.). 2009. *Alexander the Great: A new history*. Chichester: Wiley-Blackwell.

Huener, Jonathan. 2003. *Auschwitz, Poland, and the politics of commemoration, 1945–1979*. Athens: Ohio University Press.

Hilberg, Raul. 1992. *Perpetrators, victims, bystanders; The Jewish catastrophe, 1933–1945*. New York: Harper Collins.

Hilberg, Raul. 2003. *The destruction of European Jews*, 3rd ed. 3 vols. New Haven: Yale University Press.

Hilberg, Raul. 1996. *The politics of memory; The journey of a Holocaust historian*. Chicago: Ivan Dee.

Hilberg, Raul. 2001. *Sources of Holocaust research; An analysis*. Chicago: Ivan Dee.

Hirsch, Eric Donald. 1987. *Cultural literacy; What every American needs to know*. Boston: Houghton Mifflin.

Hirsch, Marianne, and Leo Spitzer. 2010. *Ghosts of home: The afterlife of Czernowitz in Jewish memory*. Berkeley: University of California Press.

Hirsch, Marianne. 2008. The generation of postmodernity. *Poetics Today* 29(1): 103–128.

Hirsch, Marianne. 2012. *The generation of postmodernity: Writing and visual culture after the Holocaust*. New York: Columbia University Press.

Hobsbawm, Eric, and Terence Ranger (eds.). 1983. *The invention of tradition.* Cambridge: Cambridge University Press.

Hoffman, Eva. 2004. *After such knowledge; Memory, history, and the legacy of the Holocaust.* New York: Public Affairs.

Homans, Peter (ed.). 2000. *Symbolic loss; The ambiguity of mourning and memory at century's end.* Charlottesville: University Press of Virginia.

Horwitz, Tony. 1999. *Confederates in the Attic; Dispatches from the unfinished civil war.* New York: Vintage.

Hoskins, Andrew. 2012. Digital network memory. In *Mediation, remediation, and the dynamics of cultural memory,* ed. Astrid Erll and Ann Rigney, 91–106. Berlin: Walter de Gruyter.

Hunt, Lynn. 1992. *The family romance of the French revolution.* Berkeley: University of California Press.

Hunt, Lynn. 2007. *Inventing human rights: A history.* New York: Norton.

Hutcheon, Linda. 1998. Irony, nostalgia, and the postmodern. University of Toronto English Department. www.library.utoronto.ca/utel/criticism/hutchinp.html

Hutton, Patrick H. 1993. *History as an art of memory.* Hanover: University Press of New England.

Hutton, Patrick H., (ed.) 2013. *Nostalgia in modern France: Bright new ideas about a melancholy subject.* Special issue of *Historical Reflections* 39(3).

Huyssen, Andreas. 1986. *After the great divide: Modernism, mass culture, postmodernism.* Bloomington: Indiana University Press.

Huyssen, Andreas. 2003. *Present pasts; Urban palimpsests and the politics of memory.* Stanford: Stanford University Press.

Huyssen, Andreas. 1995. *Twilight memories; Marking time in a culture of amnesia.* New York: Routledge.

Iggers, Georg G. 1997. *Historiography in the twentieth century; From scientific objectivity to the postmodern challenge.* Hanover: University Press of New England.

Iggers, Georg G., and Q. Edward Wang. 2008. *A global history of modern historiography.* London: Pearson.

Iggers, Georg. 1995. Historicism: The history and meaning of the term. *Journal of the History of Ideas* 56: 129–152.

Jameson, Fredric. 1989. Nostalgia for the present. *South Atlantic Quarterly* 88: 517–537.

Jameson, Fredric. 1991. *Postmodernism, or the logic of late capitalism.* Durham: Duke University Press.

Jay, Martin. 1984. *Marxism and totality; The adventures of a concept from Lukács to Habermas.* Berkeley: University of California Press.

Jay, Martin. 1973. *The dialectical imagination; A history of the Frankfurt school and the institute for social research, 1923–1950.* Boston: Little Brown.

Jencks, Charles. 1996. *What is postmodernism?* Chichester: Wiley.

Jenkins, Keith. 2003. *Refiguring history; New thoughts on an old discipline*. London: Routledge.

Jenkins, Keith. 1991. *Re-thinking history*. London: Routledge.

Jenkins, Keith. 1997a. *The postmodern history reader*. London: Routledge.

Jenkins, Keith. 1997b. *The postmodern history reader*. London: Routledge.

Jones, Marjorie. 2008. *Frances Yates and the hermetic tradition*. Lake Worth: Ibis Press.

Judt, Tony. 2010. *Ill Fares the land*. New York: Penguin.

Kammen, Michael. 1991. *Mystic chords of memory; The transformation of tradition in American culture*. New York: Knopf.

Kandel, Eric R. 2006. *In search of memory; The emergence of a new science of mind*. New York: Norton.

Kansteiner, Wulf. 2002. Finding meaning in memory: A methodological critique of collective memory studies. *History and Theory* 41: 179–197.

Kansteiner, Wulf. 2006. *In pursuit of German memory; History, television, and politics after Auschvitz*. Athens: Ohio University Press.

Kaplan, Steven Lawrence. 1995. *Farewell revolution; the historians' Feud; France 1789–1989*. Ithaca: Cornell University Press.

Kattago, Siobhan (ed.). 2015. *The Ashgate research companion to memory studies*. Surrey: Ashgate Publishing.

Kattago, Siobhan. 1997. *Ambiguous memory: The legacy of the Nazi Past in Postwar Germany*. Doctoral dissertation, The New School for Social Research.

Kelley, Donald R. 2006. *Frontiers of history; Historical inquiry in the twentieth century*. New Haven: Yale University Press.

Kelley, Donald R. (ed.). 1990. *The history of ideas; Canon and variation*. Rochester: University of Rochester Press.

Klein, Kerwin. 2000. On the emergence of memory in historical discourse. *Representations* 69: 127–150.

Knowlton, James, and Truett Cates (eds.). 1993. *Forever in the shadow of Hitler? Original documents of the Historikerstreit*. Atlantic Highlands: Humanities Press.

Koselleck, Reinhart. 1985. *Futures past; On the semantics of historical time*. Trans. K. Tribe. Cambridge, MA: MIT Press.

Kurzweil, Ray. 2005. *The singularity is near; When humans transcend biology*. New York: Penguin.

Labrousse, Ernest. 1967. *L'Histoire sociale; Sources et méthodes*. Paris: Presses Universitaires de France.

Landsberg, Alison. 2004. *Prosthetic memory; The transformation of American remembrance in the age of mass culture*. New York: Columbia University Press.

Largeaud, Jean-Marc. 2006. *Napoléon and Waterloo; la défaite glorieuse de 1815 à nos jours*. Paris: Boutique de l'Histoire.

Le Goff, Jacques, and Pierre Nora, eds. 1974. *Faire de l'histoire*. 3 vols. Paris: Gallimard.

Le Goff, Jacques. 1988. *La Nouvelle Histoire*. Paris: Editions Complexe.

Le Goff, Jacques. 1992. *History and memory*. New York: Columbia University Press.

Le Roy Ladurie, Emmanuel. 1978. *Le Territoire de l'historien*. Paris: Gallimard.

Le Roy Ladurie, Emmanuel. 1982. *Paris-Montpellier*. Paris: Gallimard.

Lefebvre, Georges. 1947. *The coming of the French revolution*. Trans. R.R. Palmer. Princeton: Princeton University Press.

Levy, Daniel, and Natan Sznaider. 2006. *The Holocaust and memory in the global age*. Trans. A. Oksiloff. Philadelphia: Temple University Press.

Levy, Daniel, and Natan Sznaider. 2010. *Human rights and memory*. University Park: Pennsylvania State University Press.

Leydesdorff, Selma, Luisa Passerini, and Paul Thompson (eds.). 1996. *Gender and memory*. New Brunswick: Transaction Publishers.

Lichtheim, George. 1966. *Marxism in modern France*. New York: Columbia University Press.

Lichtheim, George. 1961. *Marxism; An historical and critical study*. New York: Columbia University Press.

Lorenz, Chris. 2010. Unstuck in time. Or: The sudden presence of the past. In *Performing the past; Memory, history, and identity in modern Europe*, ed. Karin Tilmins, Frank van Vree, and Jay Winter, 67–102. Amsterdam: University of Amsterdam Press.

Lowenthal, David. 1996. *Possessed by the past; The heritage crusade and the spoils of history*. New York: The Free Press.

Löwy, Michael. 2005. *Reading Walter Benjamin's 'on the concept of history'*. London: Verso.

Luria, Alexander. 1976. *Cognitive development: Its cultural and social foundations*. Cambridge: Harvard University Press.

Lyotard, Jean-François. 1979. *La Condition postmoderne; rapport sur le savoir*. Paris: Editions de Minuit.

Macey, David. 1995. *The lives of Michel Foucault*. New York: Random House.

Maier, Charles S. 1988. *The unmasterable past; History, Holocaust, and German national memory*. Cambridge: Harvard University Press.

Maier, Charles S. 1980. Making time: The historiography of international relations. In *The past before us*, ed. Michael Kammen. Ithaca: Cornell University Press.

Mandrou, Robert. 1974. *Introduction à la France moderne, 1500–1640*. Paris: Albin Michel.

Marris, Michael, and Robert Paxton. 1981. *Vichy France and the Jews*. New York: Basic.

Martin, Luther H., Huck Gutman, and Patrick H. Hutton (eds.). 1988. *Technologies of the self; A seminar with Michel Foucault*. Amherst: University of Massachusetts Press.

Mazower, Mark. 1998. *Dark continent; Europe's twentieth century.* New York: Random House.

McLuhan, Marshall. 1967. The memory theatre. *Encounter* 28(3): 61–66.

McNeill, William H. 1986. *Mythistory and other essays.* Chicago: University of Chicago Press.

Miller, James. 1993. *The passion of Michel Foucault.* New York: Simon & Schuster.

Mosse, George. 1964. *The crisis of German ideology; Intellectual origins of the Third Reich.* New York: Grosset & Dunlap.

Mosse, George L. 1990. *Fallen soldiers; Reshaping the memory of the world wars.* New York: Oxford University Press.

Mosse, George L. 1975. *The nationalization of the masses; Political symbolism and mass movements in Germany from the Napoleonic wars through the Third Reich.* New York: New American Library.

Neiger, Motti, Oren Meyers, and Eyal Zandberg (eds.). 2011. *On media memory; Collective memory in a new media age.* Basingstoke: Palgrave.

Niethammer, Lutz. 1992. *Posthistoire; Has history come to an end?* London: Verso.

Nora, Pierre (ed.). 1987. *Essais d'égo-histoire.* Paris: Gallimard.

Nora, Pierre. 2013. *Esquisse d'égo histoire.* Paris: Desclée de Brouwer.

Nora, Pierre. 2011a. *Historien public.* Paris: Gallimard.

Nora, Pierre. 2002. Pour une histoire au second degré; réponse à Paul Ricoeur. *Le Débat* 122: 24–31.

Nora, Pierre, ed. 1984–1992. *Les Lieux de mémoire.* 3 vols. Paris: Gallimard.

Nora, Pierre. 2011b. *Présent, nation, mémoire.* Paris: Gallimard.

Nora, Pierre. 2002. Reasons for the current upsurge in memory. *Transit* 22: 1–8.

Nora, Pierre. 2011, November 24. Recent history and the new dangers of politicization. www.eurozine.com

Novick, Peter. 1999. *The Holocaust in American life.* Boston: Houghton Mifflin.

Novick, Peter. 1988. *That noble dream; The "objectivity question" and the American historical profession.* Cambridge: Cambridge University Press.

Olick, Jeffrey K. 2007. *The politics of regret; On collective memory and historical responsibility.* New York: Routledge.

Olick, Jeffrey K., Vered Vinitzky-Seroussi, and Daniel Levy (eds.). 2011. *The collective memory reader.* Oxford: Oxford University Press.

Olney, James. 1998. *Memory and narrative: The weave of life-writing.* Chicago: University of Chicago Press.

Ong, Walter J. 1982. *Orality and literacy; The technologizing of the word.* London: Methuen.

Ozouf, Mona, Jacques Revel, and Pierre Rosanvallon (eds.). 1994. *Histoire de la Révolution et la révolution dans l'histoire; entretiens avec François Furet.* Paris: AREHESS.

Partner, Nancy, and Sarah Foot (eds.). 2013. *The sage handbook of historical theory.* London: Sage.

Péan, Pierre. 1994. *Une Jeunesse française; François Mitterrand, 1934–1947*. Paris: Fayard.

Phillips, Mark Salber, and Gordon Schochet (eds.). 2004. *Questions of tradition*. Toronto: University of Toronto Press.

Pomian, Krystof. 1986. L'Heure des Annales. In *Les Lieux de mémoire*, ed. Pierre Nora, 2:377–429. Paris: Gallimard.

Ranger, Teremce. 1993. The invention of tradition revisited. In *Legitimacy and the state of Africa*, ed. Terence Ranger and Megan Vaughan, 62–63. London: Palgrave.

Reading, Anna. 2011. Memory and digital media: Six dynamics of the Globital memory field. In *On media memory; Collective memory in a new media age*, ed. Motti Neiger, Oren Meyers, and Eyal Zandberg, 241–252. Basingstoke: Palgrave.

Rearick, Charles. 2011. *Paris dreams, Paris memories; The city and its Mystique*. Stanford: Stanford University Press.

Ricoeur, Paul. 1980. *The contribution of French historiography to the theory of history*. Oxford: Clarendon Press.

Ricoeur, Paul. 2000. *La Mémoire, l'histoire, l'oubli*. Paris: Seuil.

Ricoeur, Paul. 2002. Mémoire: Approaches historiennes, approche philosophique. *Le Débat* 122: 41–61.

Ricoeur, Paul. 1983–1985. *Temps et récit*. 3 vols. Paris: Seuil.

Rigney, Ann. 2008. The dynamics of remembrance; Texts between monumentality and morphing. In *Cultural memory studies; An international and interdisciplinary handbook*, ed. Astrid Erll and Ansgar Nünning, 345–353. Berlin: Walter de Gruyter.

Rigney, Ann. 2010. The many afterlives of *Ivanhoe*. In *Performing the past*, ed. Karin Tilmans, Frank van Vree, and Jay Winter, 207–234. Amsterdam: University of Amsterdam Press.

Rigney, Ann. 2001. *Imperfect histories; The elusive past and the legacy of romantic historicism*. Ithaca: Cornell University Press.

Rosenfeld, Gavriel D. 2009. A looming crash or a soft landing? Forecasting the future of the memory industry. *Journal of Modern History* 81: 122–158.

Roth, Michael S., and Charles G. Salas (eds.). 2001. *Disturbing remains: Memory, history, and crisis in the twentieth century*. Los Angeles: Getty Research Institute.

Roth, Michael S. 1991. Dying of the past: Medical studies of nostalgia in nineteenth-century France. *History ad Memory* 3: 5–29.

Roth, Michael S. 1989. Remembering forgetting: *Maladies de la mémoire* in nineteenth-century France. *Representations* 26: 49–68.

Roth, Michael S. 1992. The time of nostalgia: Medicine, history, and normality in 19th-century France. *Time and Society* 1: 281–284.

Rousso, Henry. 1987. *Le Syndrome de Vichy de 1944 à nos jours*. Paris: Seuil.

Royster, Francesca T. 2003. *Becoming Cleopatra; The shifting image of an icon*. New York: Palgrave.

Schacter, Daniel L. 1996. *Searching for memory; The brain, the mind, and the past.* New York: Basic.

Schwartz, Barry. 2000. *Abraham Lincoln and the forge of national memory.* Chicago: University of Chicago Press.

Schwartz, Vanessa. 2001. Walter Benjamin for historians. *American Historical Review* 106: 1721–1743.

Seigel, Jerrold. 1990. Avoiding the subject; A Foucaultian itinerary. *Journal of the History of Ideas* 51(2): 273–299.

Shattuck, Roger. 1991. Perplexing lessons: Is there a core tradition in the humanities?". In *The hospitable canon*, ed. Virgil Nemoianu and Robert Royal, 85–96. Philadelphia: John Benjamins.

Sherman, Daniel J. 2000. *The construction of memory in interwar France.* Chicago: University of Chicago Press.

Smith, Levi. 2000. Window or mirror: The Vietnam veterans memorial and the ambiguity of remembrance. In *Symbolic loss; The ambiguity of mourning and memory at century's end*, ed. Peter Homans, 105–125. Charlottesville: University Press of Virginia.

Spiegel, Gabrielle. 2002. Memory and history: Liturgical time and historical time. *History and Theory* 41: 149–162.

Steedman, Carolyn. 2002. *Dust: The archive and cultural history.* New Brunswick: Rutgers University Press.

Tamm, Marek (ed.). 2015. *The afterlife of events; Perspectives on mnemohistory.* New York: Palgrave.

Thomson, David. 1969. *Democracy in France since 1870,* 5th ed. Oxford: Oxford University Press.

Thompson, Jenny. 2010. *Wargames; Inside the world of 20th century war reenactors.* Washington, DC: Smithsonian Books.

Tilmans, Karin, Frank van Vree, and Jay Winter (eds.). 2010. *Performing the past; Memory, history, and identity in modern Europe.* Amsterdam: Amsterdam University Press.

Tollebeek, Jo. 2004. 'Turn'd to dust and tears': Revisiting the archive. *History and Theory* 43: 237–248.

Torpey, John. 2006. *Making whole what has been smashed: On reparations politics.* Cambridge: Harvard University Press.

Vansina, Jan. 1985. *Oral tradition as history.* Madison: University of Wisconsin.

Veyne, Paul. 2010. *Foucault; His thought, his character.* Trans. J. Lloyd. Malden: Polity Press.

Vidal-Naquet, Pierre. 1992. *Assassins of memory; Essays on the denial of the Holocaust.* New York: Columbia University Press.

Warburg, Aby. 1999. *The renewal of pagan antiquity.* Intro. Kurt W. Forster. Trans. D. Britt. Los Angeles: Getty Research Institute.

Warner, Michael. 1990. *The letters of the republic; Publication and the public sphere in eighteenth-century America.* Cambridge: Harvard University Press.

White, Hayden V. 1999. *Figural realism; Studies in the mimesis effect*. Baltimore: Johns Hopkins University Press.

White, Hayden V. 1973. *Metahistory; The historical imagination in nineteenth-century Europe*. Baltimore: Johns Hopkins University Press.

Wilson, Janelle L. 2005. *Nostalgia; Sanctuary of meaning*. Lewisberg: Bucknell University Press.

Winter, Jay. 2006. *Remembering war; The great war between memory and history in the twentieth century*. New Haven: Yale University Press.

Winter, Jay. 1995. *Sites of memory, sites of mourning; The great war in European cultural history*. Cambridge: Cambridge University Press.

Yates, Frances. 1984. In *Collected essays*, ed. J.N. Hillgarth and J.B. Trapp. London: Routledge and Kegan Paul.

Yates, Frances. 1966. *The art of memory*. Chicago: University of Chicago Press.

Yerushalmi, Yosef Hayim. 2014. In *The faith of fallen Jews; Yosef Hayim Yerushalmi and the writing of Jewish history*, ed. David N. Myers and Alexander Kaye. Waltham: Brandeis University Press.

Yerushalmi, Yosef Hayim. 1996. *Zakhor; Jewish history and Jewish memory*. Seattle: University of Washington Press.

Young, James. 1993. *The texture of memory: Holocaust memorials and meaning*. New Haven: Yale University Press.

Zerubavel, Yael. 1995. *Recovered roots; Collective memory and the making of Israeli national tradition*. Chicago: University of Chicago Press.

INDEX

A

Adorno, Theodor, 8

Anderson, Benedict, on the concept of the imagined community, 51–3, 56

Anderson, Perry, on postmodernism, 13

Ankersmit, Frank, 27, 151
  on the aesthetic theory of Johan Huizinga, 163–4
  on the concept of mimesis for Erich Auerbach, 162–3
  his critique of the language theorists Richard Rorty and Jacques Derrida, 163
  on the relationship between representation and experience in historical understanding, 162–5
  on Walter Benjamin's conception of aura, 164–5

Annales school. *See* historiography, Annales scholarship

Ariès, Philippe, his history of private life, 18–19

Aristotle, *Poetics*, 162

Assmann, Aleida, 23–4, 27
  on archive *vs.* canon, 87–8
  on the concept of cultural memory, 79–81, 86, 88–92
  on cultural memory in the digital age, 91–2, 98
  on gradations of cultural memory, 88–9
  her interest in Aby Warburg, 24
  her relationship to the scholarship of Frances Yates, 79–80
  on memory as art *vs.* memory as power, 90–1, 92
  to the scholarship of Pierre Nora, 80, 81, 86–7
  on transformation of the concept of the canon, 86–7

Assmann, Jan, 23–4, 27, 79–86, 168, 191
  on the birth of literary criticism in Hellenistic culture, 85–6
  on the canon as an Egyptian invention, 83

© The Editor(s) (if applicable) and The Author(s) 2016
P.H. Hutton, *The Memory Phenomenon in Contemporary Historical Writing*, DOI 10.1057/978-1-137-49466-5

The manufacturer's authorised representative in the EU is Springer
Nature Customer Service Centre GmbH, Europaplatz 3, 69115 Heidelberg,
Germany. If you have any concerns regarding our products, please
contact ProductSafety@springernature.com

Printed and bound by CPI Group (UK) Ltd, Croydon, CR0 4YY
23/04/2026
02095595-0002